Progress in Mathematics
Vol. 13

Edited by
J. Coates and
S. Helgason

Birkhäuser
Boston · Basel · Stuttgart

Sigurdur Helgason

Topics in Harmonic Analysis on Homogeneous Spaces

1981

Birkhäuser
Boston · Basel · Stuttgart

Author:

Sigurdur Helgason
Department of Mathematics
Massachusetts Institute of Technology
Cambridge, Massachusetts 02139

Library of Congress Cataloging in Publication Data
Helgason, Sigurdur, 1927-
Topics in harmonic analysis on homogeneous spaces.
(Progress in mathematics ; v. 13)
Bibliography: p.
Includes index.
1. Harmonic analysis. 2. Homogeneous spaces. I. Title.
II. Series: Progress in mathematics (Cambridge, Mass.) ; v.13.
QA403.H35 515'.2433 81-7643
ISBN 3-7643-3051-1 AACR2

CIP-Kurtztitelaufnahme der Deutschen Bibliothek
Helgason, Sigurdur:
Topics in Harmonic Analysis on Homogeneous Spaces/
Sigurdur Helgason.-
Boston ; Basel ; Stuttgart : Birkhäuser, 1981
(Progress in Mathematics ; Vol.13)
ISBN 3-7643-3051-1
NE: GT

ISBN: 3-7643-3051-1
Printed in USA

PROGRESS IN MATHEMATICS

Already published

PM 1 Quadratic Forms in Infinite-Dimensional Vector Spaces
Herbert Gross
ISBN 3-7643-1111-8, 431 pages, $22.00, paperback

PM 2 Singularités des systèmes différentiels de Gauss-Manin
Frédéric Pham
ISBN 3-7643-3002-3, 339 pages, $18.00, paperback

PM 3 Vector Bundles on Complex Projective Spaces
C. Okonek, M. Schneider, H. Spindler
ISBN 3-7643-3000-7, 389 pages, $20.00, paperback

PM4 Complex Approximation, Proceedings, Quebec, Canada, July 3-8, 1978
Edited by Bernard Aupetit
ISBN 3-7643-3004-X, 128 pages, $10.00, paperback

PM 5 The Radon Transform
Sigurdur Helgason
ISBN 3-7643-3006-6, 202 pages, $14.00, paperback

PM 6 The Weil Representation, Maslov Index and Theta Series
Gérard Lion, Michèle Vergne
ISBN 3-7643-3007-4, 345 pages, $18.00, paperback

PM 7 Vector Bundles and Differential Equations
Proceedings, Nice, France, June 12-17, 1979
Edited by André Hirschowitz
ISBN 3-7643-3022-8, 255 pages, $14.00, paperback

PM 8 Dynamical Systems, C.I.M.E. Lectures, Bressanone, Italy, June 1978
John Guckenheimer, Jürgen Moser, Sheldon E. Newhouse
ISBN 3-7643-3024-4, 300 pages, $14.00, paperback

PM 9 Linear Algebraic Groups
T. A. Springer
ISBN 3-7643-3029-5, 304 pages, $18.00, hardcover

PM10 Ergodic Theory and Dynamical Systems I
A. Katok
ISBN 3-7643-3036-8, 352 pages, $20.00, hardcover

PM11 18th Scandinavian Congress of Mathematicians, Århus, Denmark, 1980
Edited by Erik Balslev
ISBN 3-7643-3034-6, 528 pages, $26.00, hardcover

PM12 Séminaire de Théorie des Nombres, Paris 1979-80
Edited by Marie-José Bertin
ISBN 3-7643-3035-X, 408 pages, $22.00, hardcover

CONTENTS

PREFACE

These lecture notes deal with harmonic analysis on the complete simply connected two-dimensional manifolds of constant curvature (the Euclidean plane, the two-sphere and the non-Euclidean plane). Harmonic analysis is meant here in the sense of Problem A, B and C in the Introduction. While this viewpoint is suitable for the more general class of symmetric spaces in the sense of E. Cartan we limit ourselves to the spaces above in order to make the treatment quite brief and elementary; in particular, familiarity with Lie group theory is not required. At the same time these notes are intended as an introduction to harmonic analysis on symmetric spaces. With this in mind the text is written from a general conceptual viewpoint so that most of the proofs chosen are those which generalize to symmetric spaces.

The list of references is limited to the sources quoted in the section "Bibliographical Notes" and to some expositions which are most closely related to the material and which contain more extensive bibliographies.

These notes grew out of lectures at the Canadian Mathematical Society Summer Seminar 1980. I am indebted to the participants in the seminar, particularly R. Goodman, C. Herz and N. Varopoulos, for informative discussions.

Finally, I am most grateful to Adam Koranyi for several specific suggestions.

§1. Introduction.

1. General Problems.

Let X be a locally compact Hausdorff space acted on transitively by a locally compact group G. We assume G leaves invariant a positive measure μ on X. Let T_X denote the (unitary) representation of G on the Hilbert space $L^2(X)$ defined by $(T_X(g)f)(x) = f(g^{-1}\cdot x)$ for $g \in G$, $f \in L^2(X)$, $x \in X$. By harmonic analysis on X is frequently meant the decomposition (in the sense of the so-called direct integral theory) of T_X into irreducible representations. The corresponding decomposition of any $f \in L^2(X)$ will then resemble the classical Fourier integral decomposition. The main tools in the analysis are the measure μ and the theory of operator algebras.

If $X = G/K$ is a homogeneous space of a Lie group G, K a closed subgroup, the situation changes drastically because the machinery of differential calculus becomes available. A differential operator D on X is said to be invariant under a diffeomorphism $\phi: X \longrightarrow X$ if $D(f \circ \phi) = (Df) \circ \phi$ for each $f \in C^\infty(X)$. To exploit this notion we consider the algebra $\mathbb{D}(G/K)$ of all differential operators on G/K which are invariant under the translations $\tau(g): xK \to gxK$ from G. A function on G/K which is an eigenfunction of each $D \in \mathbb{D}(G/K)$ will be called a joint eigenfunction of $\mathbb{D}(G/K)$. Given a homomorphism $\chi: \mathbb{D}(G/K) \to \mathbb{C}$ the space

$$E_\chi(X) = \{f \in C^\infty(X) : Df = \chi(D)f \quad \text{for all } D \in \mathbb{D}(G/K)\}$$

is called a joint eigenspace. Let T_χ denote the natural representation of G on $E_\chi(X)$, that is $(T_\chi(g)f)(x) = f(g^{-1}\cdot x)$. These representations are called eigenspace representations. By harmonic analysis on G/K we shall mean "answers" to the following problems:

A. Decompose "arbitrary" functions on $X = G/K$ into joint eigenfunctions of $\mathbb{D}(G/K)$.

B. Describe the joint eigenspaces $E_\chi(X)$ of $\mathbb{D}(G/K)$.

C. Determine for which χ the eigenspace representation T_χ is irreducible.

The space $C^\infty(X)$ has a standard topology described below, $E_\chi(X)$ is given the induced topology and irreducibility above means that there are no closed non-trivial invariant subspaces.

It may happen that $\mathbb{D}(G/K)$ has no non-trivial operators in which case the above considerations have no interest. However, for the symmetric spaces G/K, Problems A, B and C are reasonable and interesting. We shall discuss them in detail for $\mathbb{R}^2$, S^2 and $\mathbb{H}^2$, the three simply connected two-dimensional Riemannian manifolds of constant curvature.

2. Notation and Preliminaries.

As usual, $\mathbb{R}$ and $\mathbb{C}$ will denote the fields of real and complex numbers, respectively, and $\mathbb{Z}$ the ring of integers. The real part of a complex number c is denoted by $\mathrm{Re}(c)$. Let

$$\mathbb{R}^+ = \{t \in \mathbb{R}: t \geq 0\}, \quad \mathbb{Z}^+ = \mathbb{Z} \cap \mathbb{R}^+.$$

If X is a topological space, $C(X)$ (resp. $C_c(X)$) denotes the space of complex-valued continuous functions (resp. of compact support) on X.

Let $\mathbb{R}^n = \{x = (x_1,\dots,x_n): x_i \in \mathbb{R}\}$ and let ∂_i denote the partial derivative $\partial/\partial x_i$. If $\alpha = (\alpha_1,\dots,\alpha_n)$ is an n-tuple of integers $\alpha_i \geq 0$ we put

$$D^\alpha = \partial_1^{\alpha_1}\cdots\partial_n^{\alpha_n}, \quad x^\alpha = x_1^{\alpha_1}\cdots x_n^{\alpha_n}, \qquad |\alpha| = \alpha_1+\dots+\alpha_n.$$

Let $C^\infty(\mathbb{R}^n)$ or $\mathcal{E}(\mathbb{R}^n)$ denote the space of complex-valued C^∞ functions f on $\mathbb{R}^n$. Given $m \in \mathbb{Z}^+$ and a compact subset $K \subset \mathbb{R}^n$ let

$$(1) \qquad \|f\|_m^K = \sum_{|\alpha|\leq m} \sup_{x\in K} |(D^\alpha f)(x)|.$$

The space $\mathcal{E}(\mathbb{R}^n)$ is topologized by means of the semi-norms $\|f\|_m^K$. It is then a Frèchet space.

Let $\mathcal{D}(\mathbb{R}^n)$ or $C_c^\infty(\mathbb{R}^n)$ denote the space $C_c(\mathbb{R}^n) \cap C^\infty(\mathbb{R}^n)$. If $K \subset \mathbb{R}^n$ is compact let $\mathcal{D}_K(\mathbb{R}^n)$ denote the space of C^∞ functions with support contained in K; this space has a topology given by the norms $\| \ \|_m^K$, where $m = 0,1,2,\dots$. A linear functional T on $\mathcal{D}(\mathbb{R}^n)$ is called a <u>distribution</u> if its restriction to each $\mathcal{D}_K(\mathbb{R}^n)$ is continuous; T is said to be 0 on an open set $U \subset \mathbb{R}^n$ if $T(\phi) = 0$ for each $\phi \in \mathcal{D}(\mathbb{R}^n)$ whose support is contained in U. If V is the union of all open subsets $U_\alpha \subset \mathbb{R}^n$ on which T is 0 then a partition of unity argument shows that $T = 0$ on V. The complement of V is called the <u>support</u> of T.

If M is a manifold the notions $C^{\infty}(M) = \mathcal{E}(M)$, the topology of $\mathcal{E}(M)$, the spaces $C_c^{\infty}(M) = \mathcal{D}(M)$, $\mathcal{D}_K(M)$, distributions and their supports can be defined by means of local coordinate systems. However, we shall not need the general definitions since we only use the cases $M = U$ (open set in $\mathbb{R}^n$) and $M = \mathbb{S}^n$.

§2. The Euclidean Plane $\mathbb{R}^2$.

1. Eigenfunctions and Eigenspace Representations.

If we view $\mathbb{R}^2$ as a homogeneous space of $\mathbb{R}^2$ acting on itself by translations, Problems A, B and C boil down to ordinary Fourier analysis. In fact, the invariant differential operators are just the differential operators with constant coefficients, and the joint eigenfunctions are the constant multiples of the exponential functions. The eigenspace representations are one-dimensional, hence irreducible.

But we can also consider $\mathbb{R}^2$ as the homogeneous space $\mathbb{R}^2 = \mathbf{M}(2)/\mathbf{O}(2)$ of the group $\mathbf{M}(2)$ of all isometries of $\mathbb{R}^2$, the orthogonal group $\mathbf{O}(2)$ leaving $(0,0)$ fixed. An $\mathbf{M}(2)$-invariant differential operator on $\mathbb{R}^2$ is both translation-invariant and $\mathbf{O}(2)$-invariant. The first property implies that it has constant coefficients, the second property then implies that it is a polynomial in the Laplacian $L_{\mathbb{R}^2} = \partial^2/\partial x_1^2 + \partial^2/\partial x_2^2$. If $\lambda \in \mathbb{C}$ and ω a unit vector the function $x \longrightarrow e^{i\lambda(x,\omega)}$, where $(,)$ is the usual inner product, is an eigenfunction of $L_{\mathbb{R}^2}$ with eigenvalue $-\lambda^2$.

We write the <u>Fourier transform</u> $\tilde{f}$ of a function f on $\mathbb{R}^2$ in the form

$$(1) \qquad \tilde{f}(\lambda\omega) = \int_{\mathbb{R}^2} f(x)e^{-i\lambda(x,\omega)}dx .$$

Then the Fourier inversion formula, valid for example if $f \in \mathcal{D}(\mathbb{R}^2)$, can be written

$$f(x) = \frac{1}{(2\pi)^2} \int_{\mathbb{S}^1} \int_{\mathbb{R}^+} \tilde{f}(\lambda\omega) e^{i\lambda(x,\omega)} \lambda d\lambda \, d\omega, \tag{2}$$

and gives an explicit answer to Problem A. Here $d\omega$ denotes the circular measure on $\mathbb{S}^1$.

This formula suggests that general eigenfunctions should be obtained from the functions $e^{i\lambda(x,\omega)}$ by some kind of superposition. We shall now prove a precise result of this kind.

Given $a,b \geq 0$ let $E_{a,b}$ denote the space of holomorphic functions f on $\mathbb{C} - \{0\}$ satisfying

$$\|f\|_{a,b} = \sup_z (|f(z)| e^{-a|z| - b|z|^{-1}}) < \infty . \tag{3}$$

Then $E_{a,b}$ is a Banach space with the norm $\| \ \|_{a,b}$. Also $E_{a,b} \subset E_{a',b'}$ if and only if $a \leq a'$, $b \leq b'$ and in this case $\|f\|_{a,b} \geq \|f\|_{a',b'}$, so the injection of $E_{a,b}$ into $E_{a',b'}$ is continuous. We can give the union

$$E = \bigcup_{a,b} E_{a,b}$$

the <u>inductive limit topology</u>. This means that a fundamental system of neighborhoods of 0 is given by the <u>convex</u> sets W such that for each (a,b), $W \cap E_{a,b}$ is a neighborhood of 0 in $E_{a,b}$. We identify the members of E with their restrictions to the unit circle $\mathbb{S}^1$ and call the members of the dual space E' <u>entire functionals</u> on $\mathbb{S}^1$. Since these generalize measures it is convenient to write

$$T(f) = \int_{S^1} f(\omega)\, dT(\omega) \qquad f \in E,\ T \in E'.$$

The following result gives an answer to Problem B.

THEOREM 2.1. *The eigenfunctions of the Laplacian on* $\mathbb{R}^2$ *are precisely the harmonic functions and the functions*

$$f(x) = \int_{S^1} e^{i\lambda(x,\omega)}\, dT(\omega), \tag{4}$$

where $\lambda \in \mathbb{C} - \{0\}$ *and* T *is an entire functional on* S^1.

We note first that the right hand side of (4) is well-defined; in fact if $x = (x_1, x_2)$ the integrand is the restriction to S^1 of the function

$$z \longrightarrow \exp\left\{\tfrac{1}{2}(i\lambda)x_1(z+z^{-1}) + \tfrac{1}{2}\lambda x_2(z-z^{-1})\right\}$$

which indeed does belong to E.

Now we need to characterize the members of E in terms of their Laurent series expansions.

LEMMA 2.2. *Let* f *be holomorphic in* $\mathbb{C} - \{0\}$ *and let*

$$f(z) = \sum_n \alpha_n z^n$$

be its Laurent expansion. Then

$$f \in E_{a+\varepsilon,b+\varepsilon} \quad \underline{\text{for}}\ \underline{\text{all}} \quad \varepsilon > 0$$

<u>if and only if</u>

$$\alpha_n = 0\left(\frac{(a+\delta)^n}{n!}\right), \quad \alpha_{-n} = 0\left(\frac{(b+\delta)^n}{n!}\right), \quad n \geq 0,$$

<u>for</u> <u>each</u> $\delta > 0$.

Consider the decomposition

$$\text{(5)} \qquad f(z) = \sum_{0}^{\infty} \alpha_n z^n + \sum_{-\infty}^{-1} \alpha_n z^n = f_+(z) + f_-(z),$$

which shows that $E_{a,b} = E_{a,0} + E_{0,b}$. Thus it suffices to show that the function

$$f(z) = \sum_{0}^{\infty} \alpha_n z^n$$

belongs to $E_{a+\varepsilon,0}$ for each $\varepsilon > 0$ if and only if

$$\alpha_n = 0\left(\frac{(a+\delta)^n}{n!}\right) \qquad n \geq 0$$

for each $\delta > 0$. Assume first this last condition is satisfied. Then using Stirling's formula

$$n! = n^n e^{-n}(2\pi n)^{\frac{1}{2}} e^{\rho/12n} \qquad (0 < \rho < 1)$$

we deduce

$$\limsup_n \left(n\ |\alpha_n|^{\frac{1}{n}}\right) \leq ae.$$

Given $\delta > 0$ we have for large n

$$|\alpha_n| \leq \left(\frac{ae+\delta}{n}\right)^n .$$

We may assume this holds for all n because in order to prove $f \in E_{a+\varepsilon,0}$ we may add a polynomial to f. Then if $r = |z|$,

$$|f(z)| \leq \sum_0^\infty \alpha_n r^n \leq \sum_0^\infty \left(\frac{r(ae+\delta)}{n}\right)^n .$$

Now the function $(r(ae+\delta)/x)^x$ reaches its maximum $\exp(r(ae+\delta)/e)$ for $x = r(ae+\delta)/e$. Splitting the series into the series

$$\sum_{n \leq x_0} + \sum_{n > x_0} = S_1 + S_2,$$

where $x_0 = r(ae+2\delta)$ we have for $\varepsilon = 2\delta/e$

$$S_1 \leq r(ae + 2\delta)\exp(r(ae + \delta)/e) = 0(e^{(a+\varepsilon)r})$$

$$S_2 \leq \sum_{n>x_0} \left(\frac{r(ae+\delta)}{r(ae+2\delta)}\right)^n = 0(1),$$

which proves that $f \in E_{a+\varepsilon,0}$.

On the other hand, suppose $f \in E_{a+\varepsilon,0}$ for all $\varepsilon > 0$. Put

$$\nu = \limsup_n \left(n|\alpha_n|^{\frac{1}{n}}\right)$$

where $0 < \nu \leq \infty$. Let $0 < \delta < \nu$. Then for infinitely many n,

$$|\alpha_n| \geq \left(\frac{\nu-\delta}{n}\right)^n$$

(where $\nu - \delta$ is to be interpreted as an arbitrary large number if $\nu = \infty$). Using

$$|\alpha_n| = |f^{(n)}(0)/n!| \leq (2\pi)^{-1} \int_{|z|=r} |z|^{-n-1} |f(z)| \, |dz|$$

so

$$|\alpha_n| \leq r^{-n} \sup_{|z|=r} |f(z)|$$

we have for the sequence $r = r_n = ne/(\nu-\delta)$ (n as in the inequality above)

$$\sup_{|z|=r} |f(z)| \geq |\alpha_n| r^n \geq e^n = \exp\left(\frac{(\nu-\delta)r}{e}\right).$$

It follows that for each $\varepsilon > 0$,

$$\frac{\nu}{e} \leq a + \varepsilon$$

so $\nu \leq ae$. Given $\varepsilon > 0$ this means

$$|\alpha_n| \leq \left(\frac{ae+\varepsilon}{n}\right)^n$$

for all n sufficiently large which again by Stirling's

formula implies

$$\alpha_n = 0\left(\frac{(a+\varepsilon)^n}{n!}\right).$$

This proves the lemma.

We can also formulate the result by saying that $f(z) = \sum_0^\infty \alpha_n z^n$ belongs to $E_{a+\varepsilon,0}$ for all $\varepsilon > 0$ if and only if its Laplace transform

$$F(z) = \int_0^\infty f(t)e^{-zt}dt \qquad (\mathrm{Re}\, z > a)$$

has its Laurent series

(6) $$F(z) = \sum_0^\infty n!\alpha_n z^{-n-1}$$

convergent for $|z| > a$.

Given an entire functional $T \in E'$ we associate with it a Fourier series

(7) $$T \sim \sum_n a_n e^{in\theta},$$

where, by definition,

$$a_n = \int_{S^1} e^{-in\theta}\, dT(\theta).$$

PROPOSITION 2.3. _A series_ $\sum_n a_n e^{in\theta}$ _is the Fourier series for an entire functional if and only if_

(8) $$\sum_n |a_n| (r^{|n|}/|n|!) < \infty \quad \underline{\text{for}} \quad r \geq 0.$$

Proof. Let $T \in E'$ have the Fourier series (7). Then for each $a,b > 0$ and each $\varepsilon > 0$, T has a continuous restriction to $E_{a+\varepsilon,0}$ and to $E_{0,b+\varepsilon}$. The series $\Sigma_0^\infty a^n z^n/n!$ converges to e^{az} in the topology of $E_{a+\varepsilon,0}$. In fact, if

$$f_N(z) = \sum_{N+1}^{\infty} \frac{a^n z^n}{n!}, \quad g_N(z) = f_N(z) e^{-(a+\varepsilon)|z|}$$

then $|g_N(z)| \leq e^{-\varepsilon|z|}$ so for $R > 0$,

$$\sup_z |g_N(z)| \leq \sup_{|z| \leq R} |g_N(z)| + \sup_{|z|>R} |g_N(z)|$$

$$\leq \sup_{|z| \leq R} |f_N(z)| + \sup_{|z|>R} e^{-\varepsilon|z|}.$$

By the uniform convergence of f_N to 0 on compact sets this implies $f_N \longrightarrow 0$ in $E_{a+\varepsilon,0}$ as claimed. Consequently,

$$\int_{S^1} e^{ae^{i\theta}} dT(\theta) = \lim_{N \to 0} \sum_o^N \frac{a^n a_n}{n!},$$

Using a similar argument for $E_{o,b+\varepsilon}$, the convergence (8) follows.

On the other hand, suppose (8) holds. Then we can define a linear form T on E by

$$T(f) = \sum_n a_n \alpha_{-n}$$

if $f = \sum_n \alpha_n z^n$. By the definition of the inductive limit T is continuous provided that for each $a,b \geq 0$ the restriction $T|E_{a,b}$ is continuous. Now it is easily seen by means of the closed graph theorem that the map

$$f \in E_{a,b} \longrightarrow f_+ \in E_{a,0},$$

(in the notation of (5)), is continuous. Thus it suffices to prove that for each $a > 0$ the restriction $T|E_{a,0}$ is continuous. For $f \in E_{a,0}$ let $f(z) = \sum_0^\infty \alpha_n z^n$ denote its Taylor series and F as in (6) its Laplace transform. Let $(f_k) \subset E_{a,0}$ be a sequence converging to 0. Then

$$(9) \qquad |f_k(z)| \leq A_k \, e^{a|z|},$$

where $A_k \longrightarrow 0$. The Laplace transforms $F_k(z)$ then satisfy for $x > a$

$$(10) \qquad |F_k(x)| \leq A_k \int_0^\infty e^{(a-x)t} dt = A_k (x-a)^{-1}.$$

But since the right hand side of (9) is rotation-invariant and since the function $z \longrightarrow f(e^{i\theta}z)$ has Laplace transform $e^{-i\theta} F(e^{-i\theta}z)$ (because of (6)) the estimate in (10) implies

$$(11) \qquad |F_k(z)| \leq A_k(|z|-a)^{-1} \quad \text{for} \quad |z| > a.$$

If $\alpha_n^{(k)}$ are the Taylor coefficients of $f_k(x)$ we have

$$n!\,\alpha_n^{(k)} = (2\pi i)^{-1} \int F_k(z) z^n \, dz$$

with the integration taken for example over the circle $|z| = a+1$. But then by (11)

$$n!\,|\alpha_n^{(k)}| \leq A_k (a+1)^{n+1}$$

so

$$|T(f_k)| \leq |\sum_0^\infty a_{-n} \alpha_n^{(k)}| \leq A_k \sum_0^\infty |a_{-n}| (a+1)^{n+1} (n!)^{-1}.$$

Using our assumption (8) we deduce $T(f_k) \longrightarrow 0$, proving the continuity.

LEMMA 2.4. *Let* $\lambda \neq 0$, $m \in \mathbb{Z}$. *The solutions* f *to* $L_{\mathbb{R}^2} f = -\lambda^2 f$ *satisfying* $f(e^{i\theta} z) = e^{im\theta} f(z)$ *are the constant multiples of the function*

$$\phi_{\lambda,m}(x) = \frac{1}{2\pi} \int_{\mathbb{S}^1} e^{i\lambda(x,\omega)} \chi_m(\omega) \, d\omega \; ;$$

where $\chi_m(e^{i\theta}) = e^{im\theta}$.

Proof. We shall use the the fact that the eigenfunctions of $L_{\mathbb{R}^2}$ are analytic functions. In polar coordinates (r,θ) we have

$$L_{\mathbb{R}^2} = \frac{\partial^2}{\partial r^2} + \frac{1}{r}\frac{\partial}{\partial r} + \frac{1}{r^2}\frac{\partial^2}{\partial \theta^2}$$

so, writing $f(e^{i\theta}r) = \phi(r)e^{im\theta}$, we have

$$\frac{d^2\phi}{dr^2} + \frac{1}{r}\frac{d\phi}{dr} - \frac{m^2}{r^2}\phi = -\lambda^2\phi.$$

Although this is a second order differential equation its solutions which are smooth at the origin are all proportional. To see this we write $\phi(r) = \sum_0^\infty a_n r^n$ for small r and obtain by substitution the recursion formulas

$$m^2 a_0 = 0, \qquad (1-m^2)a_1 = 0,$$

$$((n+2)^2 - m^2)a_{n+2} = -\lambda^2 a_n, \qquad n \geq 0.$$

These equations mean that:

$a_n = 0$ for $0 \leq n < |m|$,

$a_{|m|}$ is arbitrary,

$a_{|m|+2}, a_{|m|+4}, \ldots$ are determined by $a_{|m|}$,

$a_{|m|+1} = a_{|m|+3} = \ldots = 0.$

This proves the stated proportionality. Since the function $\phi_{\lambda,n}$ is a non-zero solution the lemma is proved.

PROPOSITION 2.5. *A function* f *satisfying* $L_{\mathbb{R}^2}f = -\lambda^2 f$ *satisfies the functional equation*

$$(12) \qquad \frac{1}{2\pi}\int_0^{2\pi} f(z+e^{i\theta}w)\,d\theta = f(z)\,\phi_{\lambda,0}(w), \quad z,w \in \mathbf{C}.$$

Conversely, a continuous function f _satisfying_ (12) _is automatically of class_ C^∞ _and satisfies_ $L_{\mathbb{R}^2} f = -\lambda^2 f$.

Proof. If f is an eigenfunction, so is the function

$$F: w \longrightarrow \int_0^{2\pi} f(z+e^{i\theta}w)\,d\theta.$$

In addition F is radial so by Lemma 2.4 it equals the function $\phi_{\lambda,0}(w)$ multiplied by some factor. This factor is obtained by putting $w = 0$; (12) follows.

On the other hand, suppose f is a continuous function satisfying (12). We multiply (12) by a function $h(w) \in C_c^\infty(\mathbb{R}^2)$ and integrate with respect to w. By the invariance of dw under $\mathbb{M}(2)$ we get

$$\frac{1}{2\pi}\int_0^{2\pi} d\theta \int f(u)h(e^{-i\theta}(u-z))\,du = f(z)\int \phi_{\lambda,0}(w)h(w)\,dw$$

and this shows that f is of class C^∞.

Now apply $L = L_{\mathbb{R}^2}$ to (12) as a function of w. By its invariance under $\mathbb{M}(2)$ we get

$$\frac{1}{2\pi}\int_0^{2\pi} (Lf)(z+e^{i\theta}w)\,d\theta = f(z)(-\lambda^2)\phi_{\lambda,0}(w).$$

Putting $w = 0$ we get $Lf = -\lambda^2 f$ so the lemma is proved.

We can now prove Theorem 2.1. Let f satisfy

$L_{\mathbb{R}^2} f = -\lambda^2 f$ where $\lambda \in \mathbb{C} - \{0\}$. We expand the function $\theta \longrightarrow f(e^{i\theta}z)$ in an absolutely convergent Fourier series

$$f(e^{i\theta}z) = \sum_n c_n(z)e^{in\theta} , \tag{13}$$

where

$$c_n(z) = (2\pi)^{-1} \int_0^{2\pi} f(e^{i\theta}z)\, e^{-in\theta}\, d\theta.$$

But this is an eigenfunction of the Laplacian satisfying the homogeneity assumption of Lemma 2.4. Hence

$$c_n(z) = a_n \phi_{\lambda,n}(z), \qquad a_n \in \mathbb{C}.$$

Also, substituting $\theta = \frac{\pi}{2} - \phi$ we get

$$\phi_{\lambda,n}(r) = i^n (2\pi)^{-1} \int_0^{2\pi} e^{i\lambda r \sin\phi}\, e^{-in\phi}\, d\phi.$$

Consider now the Bessel function

$$J_n(z) = \sum_{r=0}^{\infty} (-1)^r \left(\tfrac{1}{2}z\right)^{2r+n} [r!\Gamma(n+r+1)]^{-1}$$

defined for all $n, z \in \mathbb{C}$. (For $-n \in \mathbb{Z}^+$ the first coefficients vanish). If in the integral for $\phi_{\lambda,n}$ we expand $e^{i\lambda r \sin\phi}$ in a power series and integrate term-by-term we obtain

$$\phi_{\lambda,n}(r) = i^n J_n(\lambda r).$$

On the other hand we have for $n \in \mathbb{Z}^+$

$$\Gamma(n+1)\left(\frac{1}{2}z\right)^{-n} J_n(z) = 1 + h(z),$$

where

$$|h(z)| = \left| \sum_{r=1}^{\infty} (-1)^r \left(\frac{1}{2}z\right)^{2r} \frac{1}{r!} \; \frac{\Gamma(n+1)}{\Gamma(n+r+1)} \right|$$

$$\leq \sum_{r=1}^{\infty} \left(\frac{1}{2}|z|\right)^{2r} \frac{1}{r!\,(n+1)^r} = \exp\left(\left(\frac{1}{2}|z|\right)^2/(n+1)\right) - 1$$

so

$$\lim_{n \to +\infty} n! \left(\frac{1}{2}z\right)^{-n} J_n(z) = 1.$$

Hence, the relation

$$\sum_n |a_n J_n(\lambda r)| < \infty ,$$

which follows from (13), implies

$$\sum_n |a_n| \left(r^{|n|}/|n|!\right) < \infty .$$

Thus by Proposition 2.3 there exists a $T \in E'$ such that

$$T \sim \sum a_n e^{in\theta}.$$

The formula for $\phi_{\lambda,n}$ shows that

$$e^{i\lambda(x,e^{i\theta})} = \sum_n \phi_{\lambda,n}(x) e^{-in\theta}$$

which, if $x = (x_1,x_2)$, gives the Laurent expansion

(14) $$\exp\left[\frac{1}{2}\, i\lambda\, x_1(z + z^{-1}) + \frac{1}{2}\lambda\, x_2(z - z^{-1})\right] = \sum_n \phi_{\lambda,n}(x) z^{-n}.$$

As we saw for the Taylor series for e^{az} the series (14) converges in the topology of E. Thus we can apply T to it term-by-term. This gives

$$\int_{S^1} e^{i\lambda(x,\omega)} dT(\omega) = \sum_n \phi_{\lambda,n}(x) a_n = f(x),$$

which is the desired representation of f.

On the other hand, let $T \in E'$ and consider the "integral"

$$f(x) = \int_{S^1} e^{i\lambda(x,\omega)} dT(\omega),$$

as well as the Fourier expansion

$$T \sim \Sigma\, a_n\, e^{in\theta}.$$

Applying T to the expansion (14) we get

(15) $$f(x) = \sum_n a_n\, \phi_{\lambda,n}(x);$$

by (8) and the above estimates for $J_n(\lambda r)$ this series converges uniformly on compact sets so f at least is continuous. The functions $\phi_{\lambda,n}$ all satisfy (12) and so does f because of the uniform convergence. Hence it is an eigenfunction of $L_{\mathbb{R}^2}$ and Theorem 2.1 is proved.

Problem C in §1 will now be answered with the following result.

For $\lambda \in \mathbb{C}$ let $\mathcal{E}_\lambda(\mathbb{R}^2)$ denote the eigenspace

$$\mathcal{E}_\lambda(\mathbb{R}^2) = \{f \in C^\infty(\mathbb{R}^2) : L_{\mathbb{R}^2} f = -\lambda^2 f\}$$

and let T_λ denote the natural representation of $\mathbf{M}(2)$ on $\mathcal{E}_\lambda(\mathbb{R}^2)$, that is

$$(T_\lambda(g)f)(x) = f(g^{-1}\cdot x).$$

THEOREM 2.6. *The eigenspace representation* T_λ *is irreducible if and only if* $\lambda \neq 0$.

We first have to prove a couple of lemmas. Given $F \in L^2(\mathbf{S}^1)$, consider the function

$$f(x) = \frac{1}{2\pi} \int_{\mathbf{S}^1} e^{i\lambda(x,\omega)} F(\omega)\,d\omega. \tag{16}$$

LEMMA 2.7. *Let* $\lambda \neq 0$. *Then the mapping* $F \longrightarrow f$ *defined by* (16) *is one-to-one*.

Proof. Let $p(\zeta) = p(\zeta_1, \zeta_2)$ be a polynomial and D the corresponding constant coefficient differential operator on $\mathbb{R}^2$ such that

$$D_x(e^{i(x,\zeta)}) = p(\zeta)e^{i(x,\zeta)}$$

for $\zeta \in \mathbf{C}^2$. If $f \equiv 0$ in (16) we deduce from the above equation that

$$\int_{\mathbf{S}^1} p(\lambda\omega_1,\lambda\omega_2)F(\omega)d\omega = 0.$$

Since p is arbitrary and $\lambda \neq 0$ this implies $F \equiv 0$.

For $\lambda \neq 0$ let $\mathcal{H}_\lambda$ denote the space of functions f as defined by (16); $\mathcal{H}_\lambda$ is a Hilbert space if the norm of f is defined as the L^2 norm of F on $\mathbf{S}^1$. Because of Lemma 2.7 this is well-defined.

LEMMA 2.8. <u>Let</u> $\lambda \neq 0$. <u>Then the space</u> $\mathcal{H}_\lambda$ <u>is dense in</u> $\mathcal{E}_\lambda(\mathbb{R}^2)$.

<u>Proof</u>. Let $f \in \mathcal{E}_\lambda(\mathbb{R}^2)$ and let $f^\theta(z) = f(e^{i\theta}z)$. The space $\mathcal{E}(\mathbb{R}^2)$, and hence $\mathcal{E}_\lambda(\mathbb{R}^2)$, is topologized by the semi-norms (1) §1 (with $K = \mathbf{S}^1$). The mapping $\Phi: \theta \longrightarrow f^\theta$ of $\mathbf{S}^1$ into $\mathcal{E}_\lambda(\mathbb{R}^2)$ is differentiable in the sense that for each $\theta \in \mathbf{S}^1$ the limit (in the topology of $\mathcal{E}_\lambda(\mathbb{R}^2)$)

$$\Phi'(\theta) = \lim_{h\to 0} \frac{1}{h} (\Phi(\theta+h) - \Phi(\theta))$$

exists. The series (13) can be written in the form

$$\Phi(\theta) = \sum_n a_n\phi_{\lambda,n}\, e^{in\theta}$$

so $a_n\phi_{\lambda,n}$ is the nth Fourier coefficient of the vector-valued function Φ. Since the series (13) can be differentiated at will with respect to θ we obtain for the kth derivative $\Phi^{(k)}$ of Φ

$$a_n\phi_{\lambda,n} = (in)^{-k}(2\pi)^{-1}\int_0^{2\pi} \Phi^{(k)}(\theta)e^{-in\theta}\,d\theta,$$

This gives an estimate $\|a_n\phi_{\lambda,n}\| \leq (\text{Const})\, n^{-k}$ for each seminorm $\|\ \|$ of $\mathcal{E}(\mathbb{R}^2)$; thus the series $\sum_n a_n\phi_{\lambda,n}e^{in\theta}$ converge absolutely in the topology of $\mathcal{E}(\mathbb{R}^2)$ to some limit $\Phi_0(\theta)$. But then Φ and Φ_0 will have the same Fourier coefficients so $\Phi_0 = \Phi$ and

$$\lim_{N\to\infty} \sum_{-N}^{N} a_n\phi_{\lambda,n} = \Phi(0) = f$$

in the topology of $\mathcal{E}(\mathbb{R}^2)$. This proves the lemma.

We can now prove Theorem 2.6. If $\lambda = 0$ T_λ is obviously not irreducible since the constants form a non-trivial invariant subspace. Assuming now $\lambda \neq 0$ we first prove that $\mathbf{M}(2)$ acts irreducibly on $\mathcal{H}_\lambda$. Let $V \neq 0$ be a closed invariant subspace of the Hilbert space $\mathcal{H}_\lambda$. Then there exists an $h \in V$ such that $h(0) = 1$. We write

$$h(x) = \frac{1}{2\pi}\int_{\mathbf{S}^1} e^{i\lambda(x,\omega)}H(\omega)\,d\omega.$$

Because of Lemma 2.4 the average

$$h^{\natural}(z) = \frac{1}{2\pi}\int_0^{2\pi} h(e^{i\theta}z)\,d\theta$$

is then given by

$$h^{\natural}(x) = \phi_{\lambda,0}(x) = \frac{1}{2\pi}\int_{\mathbf{S}^1} e^{i\lambda(x,\omega)}\,d\omega.$$

If f in (16) lies in the annihilator V^0 of V the functions F and H are orthogonal on $\mathbb{S}^1$. Since V^0 is $\mathbb{O}(2)$-invariant this remains true for H replaced by its average over $\mathbb{S}^1$; in other words, $\phi_{\lambda,0}$ belongs to the double annihilator $(V^0)^0$ which, by Hilbert space theory, equals V. But then since V is invariant under translations, it follows that for each $t \in \mathbb{R}^2$ the function

$$x \longrightarrow \int_{\mathbb{S}^1} e^{i\lambda(x,\omega)} e^{i\lambda(t,\omega)} d\omega$$

belongs to V. Lemma 2.7 then implies that $V^0 = \{0\}$, whence the irreducibility of $\mathbb{M}(2)$ on $\mathcal{H}_\lambda$.

Passing now to $\mathcal{E}_\lambda$ let $W \subset \mathcal{E}_\lambda$ be a closed invariant subspace. Then $W \cap \mathcal{H}_\lambda$ is an invariant subspace of $\mathcal{H}_\lambda$. Let

$$f_n(x) = \frac{1}{2\pi} \int_{\mathbb{S}^1} e^{i\lambda(x,\omega)} F_n(\omega) d\omega$$

be a sequence in $W \cap \mathcal{H}_\lambda$ converging to $f \in \mathcal{H}_\lambda$ (in the topology of $\mathcal{H}_\lambda$). With the polynomial p and the differential operator D as before we have

$$D(f_n - f)(x) = \frac{1}{2\pi} \int_0^{2\pi} e^{i\lambda(x,\omega)} p(\omega)(F_n - F)(\omega) d\omega.$$

Since $F_n \longrightarrow F$ in $L^2(\mathbb{S}^1)$ we see from Schwarz' inequality that $f_n \longrightarrow f$ in the topology of $\mathcal{E}_\lambda$. Thus $f \in W$ so $W \cap \mathcal{H}_\lambda$ is closed in $\mathcal{H}_\lambda$. Hence by the above, $W \cap \mathcal{H}_\lambda$ equals either $\mathcal{H}_\lambda$ or $\{0\}$. In the first case $W = \mathcal{E}_\lambda$, because of

Lemma 2.8. In the second case consider for each $f \in W$ the expansion (15). Each term $a_n\phi_{\lambda,n}$ is given by

$$a_n\phi_{\lambda,n}(z) = \frac{1}{2\pi}\int_0^{2\pi} f(e^{i\theta}z)e^{-in\theta}\, d\theta$$

and therefore belongs to the closed invariant subspace W as well as to $\mathcal{H}_\lambda$. It follows that $f = 0$ and the irreducibility is proved.

Consider now the exceptional value $\lambda = 0$ where the solution space $\mathcal{E}_0(\mathbb{R}^2)$ consists of the harmonic functions. We shall see that now there is a much bigger group acting, namely the conformal group (or rather its Lie algebra) and we shall show an irreducibility property for this action.

The group $\mathbf{SL}(2,\mathbb{C})$ acts transitively on the one-point compactification of the plane by means of the maps

$$g: z \longrightarrow \frac{az+b}{cz+d} \qquad z \in \mathbb{C}$$

if g is the complex matrix $\binom{ab}{cd}$ of determinant one. The Laplacian $L_{\mathbb{R}^2}$ can be written

$$L_{\mathbb{R}^2} = 4\frac{\partial^2}{\partial z\partial\bar{z}},$$

where

$$\frac{\partial}{\partial z} = \frac{1}{2}\left(\frac{\partial}{\partial x} - i\frac{\partial}{\partial y}\right), \quad \frac{\partial}{\partial\bar{z}} = \frac{1}{2}\left(\frac{\partial}{\partial x} + i\frac{\partial}{\partial y}\right).$$

If D is a differential operator on $\mathbb{R}^2$ its transform D^g by g is defined by

$$D^g : f \longrightarrow (D(f \circ g)) \circ g^{-1},$$

f being a smooth function. Then

$$\left(\left(\frac{\partial}{\partial z}\right)^g f\right)(z) = \left\{\frac{\partial}{\partial z}\left(f\left(\frac{az+b}{cz+d}\right)\right)\right\}(g^{-1}z)$$

which equals the function

$$z \longrightarrow f'\left(\frac{az+b}{cz+d}\right) \frac{1}{(cz+d)^2}$$

evaluated at $g^{-1}z$. The result is

$$\left(\frac{\partial}{\partial z}\right)^g f = (cz - a)^2 \frac{\partial f}{\partial z}, \quad \left(\frac{\partial}{\partial \bar{z}}\right)^g f = (\bar{c}\,\bar{z} - \bar{a})^2 \frac{\partial f}{\partial \bar{z}}$$

so the operator $L = L_{\mathbb{R}^2}$ satisfies

$$L^g = |cz - a|^4 L. \tag{17}$$

Consider now the Lie algebra of $\mathbf{SL}(2,\mathbb{C})$ as a six-dimensional Lie algebra $\mathfrak{sl}(2,\mathbb{C})^{\mathbb{R}}$ over $\mathbb{R}$. This Lie algebra acts continuously on $\mathcal{E}(\mathbb{R}^2)$ as follows. Let $X \in \mathfrak{sl}(2,\mathbb{C})^{\mathbb{R}}$, $u \in \mathcal{E}(\mathbb{R}^2)$ and put

$$(X^* u)\ (z) = \left\{\frac{d}{dt}\, u(\exp(-tX)\cdot z)\right\}_{t=0}, \quad z \in \mathbb{R}^2.$$

Then (17) implies that X^*u is harmonic if u is harmonic.

PROPOSITION 2.9. *The representation of* $\mathfrak{sl}(2,\mathbb{C})^{\mathbb{R}}$ *on the space* $\mathcal{E}_0(\mathbb{R}^2)$ *of harmonic functions is operator irreducible, that is the only continuous operators on* $\mathcal{E}_0(\mathbb{R}^2)$ *commuting with the action are the scalar multiples of the identity.*

Proof. Suppose $A: \mathcal{E}_0 \longrightarrow \mathcal{E}_0$ is a continuous linear mapping such that $AX^*u = X^*Au$ for all $X \in \mathfrak{sl}(2,\mathbb{C})^{\mathbb{R}}$. Taking X as

$$\begin{pmatrix} 0 & 1 \\ 0 & 0 \end{pmatrix} \quad \text{or} \quad \begin{pmatrix} 0 & i \\ 0 & 0 \end{pmatrix}$$

we see that A commutes with the partial derivatives $\partial/\partial x$, $\partial/\partial y$ and therefore also with $\partial/\partial z$ and $\partial/\partial \bar{z}$. In particular A maps the subspace $\mathcal{A} \subset \mathcal{E}_0$ of holomorphic functions and the subspace $\bar{\mathcal{A}} \subset \mathcal{E}_0$ of antiholomorphic functions into itself. Taking X as

$$\begin{pmatrix} 1 & 0 \\ 0 & -1 \end{pmatrix}$$

we see that A commutes with the operator

$$u(x,y) \longrightarrow x\frac{\partial u}{\partial x} + y\frac{\partial u}{\partial y}, \qquad u \in \mathcal{E}_0.$$

But if $u \in \mathcal{A}$ we have

$$x \frac{\partial u}{\partial x} + y \frac{\partial u}{\partial y} = z \frac{\partial u}{\partial z}$$

so A, restricted to $\mathcal{A}$, commutes with the operator $u \longrightarrow z \frac{\partial u}{\partial z}$. Putting $f_n = A(z^n)$ $(n \in \mathbf{Z}^+)$ we therefore deduce $z \frac{\partial f_n}{\partial z} = nf_n$ whence $f_n = c_n z^n$ $(c_n \in \mathbf{C})$. But since A commutes with $\partial/_{\partial z}$ we get for $n \geq 1$, $f_n' = A(nz^{n-1})$, whence $c_n = c_{n-1}$. By continuity, A is a scalar c on $\mathcal{A}$. Similarly, A is a scalar c' on $\bar{\mathcal{A}}$. But $\mathcal{A} \cap \bar{\mathcal{A}} \neq \{0\}$ so $c = c'$. Since $\mathcal{A}$ and $\bar{\mathcal{A}}$ span $\mathcal{E}_0$ the proposition follows.

2. A Paley-Wiener Theorem.

Our viewpoint of harmonic analysis on the homogeneous space $\mathbb{R}^2 = M(2)/\mathbf{O}(2)$ and more generally on $\mathbb{R}^n = M(n)/\mathbf{O}(n)$, relates functions f on $\mathbb{R}^n$ to functions ϕ on $\mathbb{R} \times \mathbf{S}^{n-1}$ via the Fourier transform

$$(18) \qquad \phi(\lambda,\omega) = \tilde{f}(\lambda\omega) = \int_{\mathbb{R}^n} f(x)e^{-i\lambda(x,\omega)}dx.$$

(We pass from $n = 2$ to general n because the proof of the result below requires a kind of an induction on the dimension). While the classical Paley-Wiener theorem characterizes the Fourier-Laplace transforms

$$(19) \qquad \tilde{f}(\zeta) = \tilde{f}(\zeta_1,\ldots,\zeta_n) = \int_{\mathbb{R}^n} f(x)e^{-i(x,\zeta)}dx, \quad f \in \mathcal{D}(\mathbb{R}^n)$$

as entire functions of exponential type in $(\zeta_1,\ldots,\zeta_n)$ it is more appropriate in the present context to ask for an intrinsic characterization of the functions $\phi(\lambda,\omega)$ in (18) as f runs through $\mathcal{D}(\mathbb{R}^n)$. (The similar problem for $f \in L^2$ offers no difficulties.) We shall now obtain such a characterization. Let $\mathrm{Im}c$ denote the imaginary part of a complex number c. A vector $a = (a_1,\ldots,a_n) \in \mathbb{C}^n$ is called <u>isotropic</u> if $a_1^2+\ldots+a_n^2 = 0$.

THEOREM 2.10. <u>The mapping</u> $f \longrightarrow \tilde{f}$ <u>maps</u> $\mathcal{D}(\mathbb{R}^n)$ <u>onto the set of functions</u> $\tilde{f}(\lambda\omega) = \phi(\lambda,\omega) \in C^\infty(\mathbb{R} \times \mathbf{S}^{n-1})$ <u>satisfying</u>:

(i) <u>For each</u> ω, <u>the function</u> $\lambda \longrightarrow \phi(\lambda,\omega)$ <u>extends to a holomorphic function on</u> $\mathbb{C}$ <u>such that for some constant</u> $A > 0$

$$\sup_{\lambda,\omega} \left| \phi(\lambda,\omega)(1+|\lambda|)^N e^{-A|\mathrm{Im}\lambda|} \right| < \infty \tag{20}$$

<u>for each</u> $N \in \mathbb{Z}$.

(ii) <u>For each</u> $k \in \mathbb{Z}^+$ <u>and each isotropic vector</u> $a \in \mathbb{C}^n$ <u>the function</u>

$$\lambda \longrightarrow \lambda^{-k} \int_{\mathbf{S}^{n-1}} \phi(\lambda,\omega)(a,\omega)^k \, d\omega$$

<u>is even and holomorphic on</u> $\mathbb{C}$.

Since the proof of this result uses several facts concerning spherical harmonics we postpone it until next section.

§3. The Sphere $\mathbf{S}^2$.

1. Spherical Harmonics.

Looking at the sphere $x_1^2 + x_2^2 + x_3^2 = 1$ as the coset space $\mathbf{O}(3)/\mathbf{O}(2)$ ($\mathbf{O}(2)$ being the subgroup leaving (0,0,1) fixed) we shall consider Problems A, B and C.

Using the spherical polar coordinates

$$x_1 = r \cos\psi \sin\theta , \; x_2 = r \sin\psi \sin\theta , \; x_3 = r \cos\theta$$

on $\mathbb{R}^3$, the Laplacian $L_{\mathbb{R}^3}$ on $\mathbb{R}^3$ takes the form

$$L_{\mathbb{R}^3} = \frac{\partial^2}{\partial r^2} + \frac{2}{r}\frac{\partial}{\partial r} + \frac{1}{r^2} L , \tag{1}$$

where

$$L = \frac{\partial^2}{\partial\theta^2} + \cot\theta \frac{\partial}{\partial\theta} + \sin^{-2}\theta \frac{\partial^2}{\partial\psi^2} . \tag{2}$$

More generally the Laplacian $L_{\mathbb{R}^n}$ $(n > 1)$ has the form

$$L_{\mathbb{R}^n} = \frac{\partial^2}{\partial r^2} + \frac{n-1}{r}\frac{\partial}{\partial r} + \frac{1}{r^2} L \tag{2a}$$

where L is the Laplacian on $\mathbf{S}^{n-1}$. It can be seen to generate the algebra of $\mathbf{O}(n)$-invariant differential operators on $\mathbf{S}^{n-1}$. Let $(s_1,\dots,s_n)$ be Cartesian coordinates of points on $\mathbf{S}^{n-1}$.

THEOREM 3.1.

(i) *The eigenspaces of L on $\mathbf{S}^{n-1}$ are of the form*

$$E_k = \text{span of } \{(a_1s_1+\dots+a_ns_n)^k : a \in \mathbb{C}^n \text{ isotropic}\}.$$

Here $k \in \mathbb{Z}^+$ *and the eigenvalue is* $-k(k+n-2)$.

(ii) *Each eigenspace representation is irreducible.*

(iii) $L^2(S^{n-1}) = \sum_0^\infty E_k$ (orthogonal Hilbert-space decomposition).

Proof. Let P_k denote the space of homogeneous polynomial functions $p(x_1,\ldots,x_n)$ of degree k on $\mathbb{R}^n$ and $H_k \subset P_k$ the subspace of harmonic polynomials. Consider the bilinear form $<,>$ on $P = \sum_0^\infty P_k$ given by

$$<p,q> = \left(p\left(\frac{\partial}{\partial x_1},\ldots,\frac{\partial}{\partial x_n}\right)q\right)(0).$$

It is clearly nondegenerate. We also see that if $p,q,r \in P$ and if

$$\partial(p) = p\left(\frac{\partial}{\partial x_1},\ldots,\frac{\partial}{\partial x_n}\right)$$

we have

$$<p,qr> = (\partial(qr)p)(0) = (\partial(r)\partial(q)p)(0) = <\partial(q)p,r>$$

so the operators $r \longrightarrow qr$ and $r \longrightarrow \partial(q)r$ are adjoint operators. In particular, if $p \in P_{k-2}$, $q = x_1^2 + \ldots + x_n^2$ we have

$$<qp,h> = <p,\partial(q)h>$$

so that H_k is the orthogonal complement of qP_{k-2} in P_k; hence

(3) $$P_k = qP_{k-2} + H_k$$

and by iteration

(4) $$P_k = H_k + |x|^2 H_{k-2} + \ldots + |x|^{2m} H_{k-2m}, \quad m = [\tfrac{1}{2}k].$$

On the other hand, if $c = (c_1, \ldots, c_n) \in \mathbb{C}^n$ such that $(c,c) = c_1^2 + \ldots + c_n^2 = 0$ then the polynomial $h_c(x) = (c_1x_1 + \ldots + c_nx_n)^k \in H_k$. Let H_k^0 be the span of these polynomials $h_c(x)$ for $(c,c) = 0$. Let $h \in H_k$ be orthogonal to H_k^0. A simple computation shows for $p \in P_k$

$$\partial(p)(c_1x_1 + \ldots + c_nx_n)^m = m(m-1)\ldots(m-k+1)p(c)(c_1x_1 + \ldots + c_nx_n)^{m-k}$$

so in particular

$$h(c) = 0 \quad \text{if} \quad (c,c) = 0.$$

In other words, h vanishes identically on the variety $\{c \in \mathbb{C}^n : (c,c) = 0\}$. By Hilbert's Nullstellensatz some power h^m belongs to the ideal $(x_1^2 + \ldots + x_n^2)P$. If $n \geq 3$, this is a prime ideal so h itself is divisible by $x_1^2 + \ldots + x_n^2$ whence by (3), $h = 0$. Consequently, if $n \geq 3$

(5) $$H_k = \text{span of } \{(c_1x_1 + \ldots + c_nx_n)^k : (c,c) = 0\}.$$

This holds also for $n = 2$; in fact we can represent a real harmonic polynomial $u(x_1, x_2)$ as the real part of a holomorphic fuction $f(x_1 + ix_2) = u(x_1,x_2) + iv(x_1,x_2)$. By the Cauchy-Riemann equations v and therefore f is a polynomial. Hence u is a linear combination of powers $(x_1 \pm ix_2)^k$.

By definition, E_k consists of the restrictions $H_k|\mathbf{S}^{n-1}$. It follows from (2a) that each $h \in E_k$ satisfies $Lh = -k(k+n-2)h$. The eigenvalues $-k(k+n-2)$ being different for different k, the spaces E_k are mutually orthogonal. By the Stone-Weierstrass theorem the algebraic sum $\sum_0^\infty P_k|\mathbf{S}^{n-1}$, which by (4) equals $\sum_0^\infty E_k$, is dense in $C(\mathbf{S}^{n-1})$ in the uniform norm; this implies (iii). Let $S_{km}(1 \leq m \leq d(k))$ be an orthonormal basis of E_k. Let f be any eigenfunction of L, $Lf = cf(c \in \mathbf{C})$. We have the expansion

$$f \sim \sum_{k,m} a_{km} S_{km}, \qquad a_{km} = \langle f, S_{km} \rangle$$

where $\langle , \rangle$ is the inner product on $L^2(\mathbf{S}^{n-1})$. But then

$$c \sum_{k,m} a_{km} S_{km} \sim cf = Lf \sim \sum_{m,k} -k(k+n-2) \; a_{km} S_{km}$$

so c is one of the eigenvalues $-k_0(k_0+n-2)$ and E_{k_0} is the corresponding eigenspace. This proves (i). For (ii) suppose $E_k = E' \oplus E''$ is an orthogonal decomposition of E_k into two nonzero invariant subspaces. By the $\mathbf{O}(n)$-invariance each of the spaces E', E'' contains a function which equals 1 at the point $(0,\ldots,0,1)$. By averaging over $\mathbf{O}(n-1)$ we would get $\mathbf{O}(n-1)$-invariant functions $\phi' \in E'$, $\phi'' \in E''$ with $\phi'(0,\ldots,0,1) = \phi''(0,\ldots,0,1) = 1$.

The functions ϕ' and ϕ'' only depend on the distance θ from $(0,\ldots,0,1)$. On such functions, L has the form

$$\frac{d^2}{d\theta^2} + (n-2)\cot\theta \ \frac{d}{d\theta}$$

(cf.(2)). Thus both ϕ' and ϕ'' satisfy the singular differential equation

$$\frac{d\phi^2}{d\theta^2} + (n-2)\cot\theta \ \frac{d\phi}{d\theta} = -k(k+n-2)\phi$$

and have a power series expansion $\phi(\theta) = \sum_{o}^{\infty} a_m \sin^m \theta$, at least for θ sufficiently small. By direct substitution the equation gives the formulas

$$a_1 = 0, \ (m+2)(m+n-1)a_{m+2} = \left[m(m+n-2)-k(k+n-2)\right]a_m.$$

This shows that ϕ is determined up to a constant factor. Hence ϕ' and ϕ'' coincide so Theorem 3.1 is proved.

Remark. The eigenfunctions of L on S^{n-1} are called spherical harmonics. The name is derived from the fact that they coincide with the set of restrictions $\cup_{k\geq 0}(H_k|S^{n-1})$.

2. Proof of Theorem 2.10.

We first prove that conditions (i) and (ii) are satisfied

for each $\tilde{f}$ if $f \in \mathcal{D}(\mathbb{R}^n)$. First note that (18) §2 can be written

$$(6) \qquad \phi(\lambda,\omega) = \tilde{f}(\lambda\omega) = \int_{\mathbb{R}} \hat{f}(\omega,\rho)e^{-i\lambda\rho}\,d\rho,$$

where $\hat{f}$ is the integral of f over the hyperplane $(x,\omega) = \rho$. Since the function $\rho \longrightarrow \hat{f}(\omega,\rho)$ belongs to $\mathcal{D}(\mathbb{R})$ property (i) follows immediately. Also (18) §2 implies that $d^k\tilde{f}/d\lambda^k$ at $\lambda = 0$ is a homogeneous k^{th} degree polynomial in $(\omega_1,\ldots,\omega_n)$ which according to (4) can be expressed as a harmonic polynomial of degree $\leq k$. Expanding the function $\lambda \longrightarrow \phi(\lambda,\omega) = \tilde{f}(\lambda\omega)$ in a convergent Taylor series around $\lambda = 0$ and integrating against $(a,\omega)^k$, property (ii) follows since E_k and E_ℓ are orthogonal for $k > \ell$.

Conversely, suppose ϕ satisfies (i) and (ii) and define $f \in C^\infty(\mathbb{R}^n)$ by

$$(7) \qquad f(x) = (2\pi)^{-n} \int_{\mathbb{R}^+ \times \mathbf{S}^{n-1}} \phi(\lambda,\omega)e^{i\lambda(x,\omega)}\lambda^{n-1}d\lambda\, d\omega.$$

Generalizing expansion (13) §2 we have the convergent expansion

$$(8) \qquad f(x) = \sum_\delta f_\delta(x),$$

where

$$(9) \qquad f_\delta(x) = d(\delta) \int_{\mathbf{O}(n)} f(u\cdot x)\chi_\delta(u^{-1})du.$$

Here du is the normalized Haar measure on the orthogonal group $\mathbf{O}(n)$, and δ runs through the unitary irreducible representations of $\mathbf{O}(n)$; χ_δ and $d(\delta)$ denote the character and the degree of δ, respectively. Formula (8) follows from the Peter-Weyl theorem (cf. Weil [1940])

$$(10) \qquad F(u) = \sum_\delta d(\delta)F*\chi_\delta(u), \qquad u \in U,$$

where

$$F*\chi_\delta(u) = \int_{\mathbf{O}(n)} F(uk)\chi_\delta(k^{-1})dk,$$

used on the function $F(u) = f(u\cdot x)$.

Going back to (7) we expand $\phi(\lambda,\omega)$ according to Theorem 3.1 (ii) into spherical harmonics,

$$(11) \qquad \phi(\lambda,\omega) = \sum_{o\leq k} \sum_{1\leq m\leq d(\delta)} \phi_{km}(\lambda)\ S_{km}(\omega).$$

Here S_{km} is the basis of E_k as above and

$$\phi_{km}(\lambda) = \int_{\mathbf{S}^{n-1}} \phi(\lambda,\omega)\ \overline{S}_{km}(\omega)\ d\omega.$$

Replacing ϕ by $L_\omega^\rho(\phi(\lambda,\omega))$ we see that $\Sigma_{k,m}|\phi_{km}(\lambda)|^2k^{2\rho}$ converges; later we shall verify that $S_{km}(\omega)$ grows at most like a power of k so (11) converges absolutely and uniformly in ω.

Putting

$$f_{km}(x) = (2\pi)^{-n} \int_{\mathbb{R}^+ \times \mathbf{S}^{n-1}} \phi_{km}(\lambda) S_{km}(\omega) e^{i\lambda(x,\omega)} \lambda^{n-1} d\lambda d\omega$$

we would expect from (7) and (11) the convergent expansion

$$(12) \qquad f(x) = \sum_{k,m} f_{km}(x).$$

However, the term-by-term integration over $\mathbb{R}^+ \times \mathbf{S}^{n-1}$ of the series (11) is not immediately justified because, although $\phi_{km}(\lambda)$ is rapidly decreasing in k and in λ it is not so clear at this point that it is rapidly decreasing in (k,λ). Instead we shall derive (12) directly from (8) a bit later.

Assuming (12) our problem is to prove for each k,m

$$(13) \qquad \int_{\mathbb{R}^+ \times \mathbf{S}^{n-1}} \phi_{km}(\lambda) S_{km}(\omega) e^{i\lambda(x,\omega)} \lambda^{n-1} d\lambda \, d\omega = 0$$

for $|x| > A$. Since $\phi_{km}(\lambda)$ is the integral of $\phi(\lambda,\omega)$ against a linear combination of polynomials $(a,\omega)^k$ where $a \in \mathbf{C}^n$ is isotropic it follows from (i) and (ii) that $\phi_{km}(\lambda)\lambda^{-k}$ is even and holomorphic and

$$(14) \qquad \left|\phi_{km}(\lambda)\right| \leq C_N (1+|\lambda|)^{-N} e^{A|\mathrm{Im}\lambda|} .$$

Now we need the following classical formula (generalizing (13a) §2)

$$\int_{\mathbf{S}^{n-1}} e^{i\lambda(\eta,\omega)} S_{km}(\omega)\,d\omega = c_{n,k} S_{km}(\eta) \frac{J_{k+\frac{1}{2}n-1}(\lambda)}{\lambda^{\frac{1}{2}n-1}}, \tag{15}$$

where $\eta \in \mathbf{S}^{n-1}$ and $c_{n,k} = (2\pi)^{\frac{n}{2}} i^k$ (cf. e.g. Bochner [1955], p.37, Erdèlyi [1953], Vol. II, p.247 and Vilenkin [1968], p. 554). We shall indicate a group-theoretic proof later. Since the assumptions (i) and (ii) are invariant under rotations of ω the problem (13) reduces to proving

$$\int_0^\infty \phi_{km}(\lambda)(\lambda r)^{1-\frac{1}{2}n} J_{k+\frac{1}{2}n-1}(\lambda r)\lambda^{n-1}\,d\lambda = 0 \tag{16}$$

for $r > A$.

First let us assume $\phi(\lambda,\omega) = \phi(\lambda)$ is independent of ω. We "extend" ϕ to an invariant holomorphic function Φ of n variables by putting

$$\Phi(\zeta) = \Phi(\zeta_1,\cdots,\zeta_n) = \phi(\lambda)$$

if $\lambda^2 = \zeta_1^2 + \cdots + \zeta_n^2$. Since ϕ is even this is possible. Writing

$$\lambda = \mu + i\nu, \qquad \zeta = \xi + i\eta, \qquad \xi,\eta \in \mathbb{R}^n$$

we have

$$\mu^2 - \nu^2 = |\xi|^2 - |\eta|^2, \quad \mu^2\nu^2 = (\xi\cdot\eta)^2, \tag{17}$$

whence

$$|\lambda|^4 = (|\xi|^2 - |\eta|^2)^2 + 4(\xi\cdot\eta)^2 \tag{18}$$

and

$$2|\mathrm{Im}\lambda|^2 = |\eta|^2 - |\xi|^2 + \left[(|\xi|^2 - |\eta|^2)^2 + 4(\xi\cdot\eta)^2\right]^{\frac{1}{2}}. \tag{19}$$

In particular, since $|(\xi\cdot\eta)| \leq |\xi||\eta|$ we have

$$|\mathrm{Im}\lambda| \leq |\eta|. \tag{20}$$

Now ϕ satisfies for each $N \in \mathbb{Z}^+$

$$|\phi(\lambda)| \leq C_N \, (1+|\lambda|)^{-N} \, e^{A|\mathrm{Im}\lambda|} \tag{21}$$

and (7) can be written

$$f(x) = (2\pi)^{-n} \int_{\mathbb{R}^n} \Phi(\xi_1,\ldots,\xi_n) e^{i(x,\xi)} d\xi. \tag{22}$$

By a suitable variation of the classical Paley-Wiener theorem argument (cf. Hörmander [1963]) we shall prove $f \in \mathcal{D}(\mathbb{R}^n)$.

If $\eta \in \mathbb{R}^n$ is any fixed vector we can by Cauchy's theorem shift the integration in (22) into the complex domain giving

$$(23) \qquad f(x) = (2\pi)^{-n} \int_{\mathbb{R}^n} \Phi(\xi+i\eta) e^{i(x,\xi+i\eta)} d\xi.$$

In fact, (18) and (21) show that, η being fixed, $\Phi(\xi+it\eta)$ is rapidly decreasing in ξ, uniformly for $0 \leq t \leq 1$. Also, by (20), and (18), we have

$$\begin{aligned} |\Phi(\xi+i\eta)| &\leq C_N \ (1+|\lambda|)^{-N} \ e^{A|\eta|} \\ &\leq C_N \ e^{A|\eta|} \left(1+\left||\xi|^2-|\eta|^2\right|^{\frac{1}{2}}\right)^{-N}. \end{aligned}$$

Taking $N = n + 1$ we have

$$\int_{\mathbb{R}^n} \left(1+\left||\xi|^2-|\eta|^2\right|^{\frac{1}{2}}\right)^{-N} d\xi \leq a|\eta|^n+b$$

where a and b are constants, as we see by breaking the integral up into the regions $|\xi| \leq 2|\eta|$ and $|\xi| > 2|\eta|$. Thus (23) implies

$$(24) \qquad |f(x)| \leq (2\pi)^{-n} C_{n+1} (a|\eta|^n+b) e^{A|\eta|-(x,\eta)},$$

valid for all $x,\eta \in \mathbb{R}^n$. Now fix x with $|x| > A$ and put $\eta = tx$ in (24).

Since

$$(a|x|^n t^n + b)e^{t|x|(A-|x|)} \longrightarrow 0$$

as $t \longrightarrow +\infty$ we obtain

$$(25) \qquad f(x) = 0 \quad \text{for} \quad |x| > A.$$

Consider now the general $\phi(\lambda,\omega)$ and its expansion (11). We recall that $\phi_{km}(\lambda)\lambda^{-k}$ is even and holomorphic and ϕ_{km} satisfies (14). We now write the expression in (16) as r^k times

$$(26) \qquad r^{-(\frac{1}{2}n+k)+1}\int_0^\infty (\phi_{km}(\lambda)\lambda^{-k})J_{k+\frac{1}{2}n-1}(\lambda r)\lambda^{\frac{1}{2}n+k}d\lambda.$$

But as a special case of (15) we have

$$\int_{S^{n-1+2k}} e^{i\lambda(x,\omega)}d\omega = c\,\frac{J_{k+\frac{1}{2}n-1}(\lambda r)}{(\lambda r)^{\frac{1}{2}n+k-1}},$$

where $r = |x|$ and c a constant. Consequently, the integral (26) is up to a constant factor the Fourier transform in $\mathbb{R}^{n+2k}$ of the radial function

$$\Phi_{km}(\xi_1,\cdots,\xi_{n+2k}) = \phi_{km}(|\xi|)|\xi|^{-k}, \quad \xi \in \mathbb{R}^{n+2k}.$$

Letting $\phi_{km}(\lambda)\lambda^{-k}$ play the role of $\phi(\lambda)$ above the function $\Phi(\zeta_1,\cdots,\zeta_{n+2k})$ restricts to Φ_{km} on $\mathbb{R}^{n+2k}$. By the conclusion (25) we therefore find that the integral (26) vanishes for $r > A$ as desired.

It remains to justify the expansion (12), the indicated growth of $S_{k,m}$ and to sketch a proof of (15) by means of representation theory.

Let U denote the orthogonal group $\mathbf{O}(n)$ and M the subgroup of U leaving the point $e_n = (0,\ldots,0,1) \in \mathbf{S}^{n-1}$ fixed. The mapping $uM \longrightarrow ue_n$ identifies $U/_M$ with $\mathbf{S}^{n-1}$. Let du be the normalized Haar measure on U. Then

$$(27) \qquad \int_U f(u\cdot e_n)du = \frac{1}{A(\mathbf{S}^{n-1})}\int_{\mathbf{S}^{n-1}} f(\omega)d\omega, \quad f \in C(\mathbf{S}^{n-1}),$$

$A(\mathbf{S}^{n-1})$ denoting the area of $\mathbf{S}^{n-1}$.

Let δ be an irreducible unitary representation of U on a Hilbert space V_δ of dimension $d(\delta) < \infty$. Let $\langle , \rangle$ denote the inner product on V_δ and V_δ^M the space of vectors fixed under M. We assume $V_\delta^M \neq 0$. Then it is known that $\dim V_\delta^M = 1$. (If v_1, v_2 are orthonormal vectors in V_δ^M the functions $uM \longrightarrow \langle v_i, \delta(u)v_i\rangle$ for $i = 1,2$ would only depend on the distance θ from e_n and would be eigenfunctions of the operator $d^2/d\theta^2 + (n-2)\cot\theta \, d/d\theta$ with the same eigenvalue. By the end of the proof of Theorem 3.1 these functions must be proportional, yet by the Schur orthogonality relations they are orthogonal).

Fix a unit vector $v \in V_\delta^M$, and let E_δ denote the space of functions $uM \longrightarrow \langle w, \delta(u)v\rangle$ on $U/_M$ as w runs through V_δ. Then the Peter-Weyl theory on $U/_M$ (cf. Weil [1940] p.81) implies $L^2(\mathbf{S}^{n-1}) = \sum_\delta E_\delta$ (orthogonal Hilbert space sum). It is clear that this coincides with the decomposition in Theorem 3.1 (iii), i.e.

$$E_\delta = E_k \qquad \text{for a suitable} \quad k \in \mathbb{Z}^+.$$

The corresponding eigenvalue is $-k(k+n-2)$. Moreover, if (w_i) is an orthonormal basis of V_δ and $w_1 = v$ the Schur orthogonality relations show that the functions

$$(28) \qquad d(\delta)^{\frac{1}{2}}\langle w_m, \delta(u)v\rangle \qquad (1 \leq m \leq d(\delta))$$

are orthonormal and therefore can serve as $\bar{S}_{km}(uM)$ where $S_{km}(1 \leq m \leq d(k))$ is the basis above. From (3) we see that $d(k)$ is bounded by a fixed power of k; the same is therefore the case for the functions (28) and the S_{km} as indicated before.

Let

$$(29) \qquad \Phi_{\lambda,\delta}(x) = \int_U e^{i\lambda(x,ue_n)}\langle v, \delta(u)v\rangle du, \qquad x \in \mathbb{R}^n.$$

LEMMA 3.2. <u>Let</u> $w \in V_\delta$. <u>Then, if</u> $x = \ell re_n$, $\ell \in U$,

$$(30) \qquad \int_U e^{i\lambda(x,ue_n)}\langle w, \delta(u)v\rangle du = \langle w, \delta(\ell)v\rangle \; \Phi_{\lambda,\delta}(re_n).$$

<u>Proof</u>. Define the map $F\colon \mathbb{R}^n \longrightarrow V_\delta$ by

$$F(x) = \int_U e^{-i\bar{\lambda}(x,u\cdot e_n)}\delta(u)v \, du.$$

Then the left hand side of (30) equals $\langle w, F(x)\rangle$. Also $F(u\cdot x) = \delta(u)F(x)$ so $F(re_n) \in V_\delta^M$. It follows that

$$F(re_n) = (\Phi_{\lambda,\delta}(re_n))^- v$$

so

$$\langle w, F(\ell\cdot re_n)\rangle = \langle w, \delta(\ell)F(re_n)\rangle$$

$$= \langle w, \delta(\ell)v\rangle \Phi_{\lambda,\delta}(re_n),$$

proving the lemma.

It remains to compute $\Phi_{\lambda,\delta}(x)$. We have of course

$$L_{\mathbb{R}^n}\Phi_{\lambda,\delta} = -\lambda^2\ \Phi_{\lambda,\delta} \tag{31}$$

and we shall now use (2a) to separate the variables. Since $\Phi_{\lambda,\delta}$ is M-invariant it is not hard to see that $\Phi_{\lambda,\delta}(\ell\cdot e_n) = \Phi_{\lambda,\delta}(\ell^{-1}e_n)$. Hence

$$\Phi_{\lambda,\delta}(r\ell e_n) = \Phi_{\lambda,\delta}(r\ell^{-1}e_n) = \int_U e^{i\lambda r(e_n,\ell ue_n)}\langle v, \delta(u)v\rangle du$$

so

$$\Phi_{\lambda,\delta}(r\ell e_n) = \int_U e^{i\lambda r(e_n,ue_n)}\langle \delta(\ell)v, \delta(u)v\rangle du.$$

Thus for a fixed r, the function $\ell M \longrightarrow \Phi_{\lambda,\delta}(r\ell e_n)$ on

S^{n-1} is an eigenfunction of L with eigenvalue $-k(k+n-2)$. Combining this with (31) and (2a) we deduce that the function $\psi(r) = \Phi_{\lambda,\delta}(re_n)$ satisfies the differential equation

$$\psi'' + \frac{n-1}{r}\psi' - \frac{k(k+n-2)}{r^2}\psi = -\lambda^2\psi. \tag{32}$$

But the constant multiples of the function $J_{k+\frac{1}{2}n-1}(\lambda r)/(\lambda r)^{\frac{1}{2}n-1}$ are the smooth solutions to this differential equation. This, combined with Lemma 3.2, gives formula (15) up to a constant factor which we shall not determine.

Finally, we give a justification of (12). Let δ be a unitary irreducible representation of U on V_δ. For $f \in C^\infty(\mathbb{R}^n)$ put

$$f^\delta(x) = d(\delta)\int_U f(u\cdot x)\delta(u^{-1})\,du. \tag{33}$$

Then f^δ is a C^∞ function with values in $\mathrm{Hom}(V_\delta, V_\delta)$ and

$$f^\delta(u\cdot x) = \delta(u)f^\delta(x). \tag{34}$$

This implies that $f^\delta \equiv 0$ unless $V_\delta^M \neq 0$. In fact $f^\delta \not\equiv 0$ implies by (34) $f^\delta(re_1) \neq 0$ for some r so $f^\delta(re_1)v \neq 0$ for some $v \in V_\delta$. But then

$$\delta(m)f^\delta(re_1)v = f^\delta(mre_1)v = f^\delta(re_1)v$$

for all $m \in M$.

Now let f be given by (7). Then writing $\omega = u\cdot e_n$ $(u \in U)$ we find

$$f^{\delta}(x) = (2\pi)^{-n} \int_{\mathbb{R}^+} \left[\int_U \delta(u) e^{i\lambda(x,uM)} du\right] \phi_\delta(\lambda)\lambda^{n-1} d\lambda,$$

where

(35) $$\phi_\delta(\lambda) = d(\delta) \int_U \phi(\lambda, uM)\delta(u^{-1}) du.$$

Clearly $\phi_\delta(\lambda)$ maps V_δ into V_δ^M so $\phi_\delta(\lambda)w_m = \langle\phi_\delta(\lambda)w_m, v\rangle v$. Hence

(36) $$\begin{aligned}\phi_\delta(\lambda)w_m &= d(\delta)\left[\int_U \phi(\lambda, uM)\langle w_m, \delta(u)v\rangle du\right] v \\ &= d(\delta)^{\frac{1}{2}} \left[\int_{\mathbf{S}^{n-1}} \phi(\lambda,\omega)\ \overline{S}_{km}(\omega) d\omega\right] v\end{aligned}$$

so

(37) $$\phi_\delta(\lambda)w_m = d(\delta)^{\frac{1}{2}}\ \phi_{km}(\lambda)v.$$

Then (34) gives

(38) $$\langle f^{\delta}(x)w_m, w_m\rangle = (2\pi)^{-n} \int_{R^+\times\mathbf{S}^{n-1}} \phi_{km}(\lambda) S_{km}(\omega) e^{i\lambda(x,\omega)} \lambda^{n-1} d\lambda d\omega,$$

which equals the function $f_{km}(x)$ defined earlier. Thus we see that the desired expansion (12) follows from (8). This concludes the proof of Theorem 2.10.

§4. The Non-Euclidean Plane.

From the history of the non-Euclidean geometry which among others involves the names Euclid, Proclus, Saccheri, Legendre, Gauss, Schweikart, Taurinus, W. Bolyai, J. Bolyai, Lobachevsky, Beltrami, Poincaré, Klein, Hilbert, we list a few quotations.

1. Wolfgang Bolyai, in a letter to his son Janos, writes about his efforts to prove Euclid's parallelaxiom:

"Ich hatte mir vorgenommen mich für die Wahrheit aufzuopfern;ich wäre bereit gewesen zum Märtyrer zu werden damit ich nur die Geometrie von diesem Makel gereinigt dem menschlichen Geschlecht übergeben könnte. Schauderhafte riesige Arbeiten habe ich vollbracht...aber keine vollkommene Befriedigung habe ich gefunden.... . Ich bin zurückgekehrt, als ich durchschaut habe, dass man den Boden dieser Nacht von der Erde aus nicht erreichen kann, ohne Trost, mich selbst und das ganze menschliche Geschlecht bedauernd. Lerne an meinen Beispiel; indem ich die Parallelen kennen wollte, blieb ich unwissend, diese haben mir all' die Blumen meines Lebens und meiner Zeit weggenommen."

2. Janos Bolyai to his father Wolfgang:

"Aus Nichts habe ich eine neue andere Welt geschaffen."

3. Wolfgang to Janos:

"Manche Dinge haben gleichsam eine Epoche, wo sie dann an mehreren Orten aufgefunden werden, gleichwie im Frühjahr die Veilchen mehrwärts ans Licht hervorkommen."

4. Gauss to Wolfgang Bolyai:

"Die Arbeit Deines Sohnes loben hiesse mich selbst loben: denn der ganze Inhalt der Schrift, der Weg, den Dein Sohn eingeschlagen hat, und die Resultate, zu denen er geführt ist, kommen fast durchgehends mit meinen eigenen, zum Theile schon seit 30-35 Jahren angestellten Meditationen überein... . Höchst erfreulich ist es mir, dass gerade der Sohn meines alten Freundes es ist, der mir auf eine so merkwürdige Art zuvorgekommen ist.

5. *"Die neue Geometrie .. eröffnet ein neues weiteres Feld für die Anwendungen von Geometrie und Analysis auf einander."*

Lobachevsky.

6. *"Ein Student der Mathematik war anwesend... Er hatte die Behauptung aufgestellt dass man durch einen Punkt mehr als eine Parallele zu einer Geraden ziehen könne. ... Nun bewies er es so schlagend, dass alle taten, als hätten sie es verstanden."*

Thomas Mann in *"Der kleine Herr Friedemann"*.

1. Problem and Results.

Let D be the open disk $|z| < 1$ in $\mathbb{R}^2$ with the usual differentiable structure but given the Riemannian structure

(1) $$\langle u,v\rangle_z = \frac{(u,v)}{(1-|z|^2)^2} .$$

Here u and v are any tangent vectors at $z \in D$ and $(,)$

denotes the usual inner product on $\mathbb{R}^2$. This Riemannian manifold is usually called the Poincaré model of the non-Euclidean plane. We shall now discuss some of its geometric properties. Since

$$\frac{\langle u,v\rangle^2}{\langle u,u\rangle \langle v,v\rangle} = \frac{(u,v)^2}{(u,u)\ (v,v)}$$

the angle between vectors in this Riemannian structure coincides with the Euclidean angle.

The length of a curve $\gamma(t)$ $(\alpha \le t \le \beta)$ in D is, according to the general definition for a Riemannian manifold, defined by

$$L(\gamma) = \int_\alpha^\beta \langle \gamma'(t), \gamma'(t)\rangle^{\frac{1}{2}}\, dt$$

and the distance between any two points $z, w \in D$ defined by

$$d(z,w) = \inf_\gamma L(\gamma),$$

the infimum taken over all curves joining z and w. If $\gamma(t) = (x(t), y(t))$ and $s(\tau)$ the arc length of the segment $\gamma(t)$ $(\alpha \le t \le \tau)$ we get from (1)

$$\left(\frac{ds}{d\tau}\right)^2 = \left(1-x(\tau)^2 - y(\tau)^2\right)^{-2} \left[\left(\frac{dx}{d\tau}\right)^2 + \left(\frac{dy}{d\tau}\right)^2\right],$$

which is the meaning of the customary terminology

$$ds^2 = \frac{dx^2 + dy^2}{(1-x^2-y^2)^2} \quad . \tag{2}$$

In particular, if $\gamma(\alpha) = o$ (the origin), $\gamma(\beta)=x$ $(0 \leq x < 1)$ and we denote by γ_0 the line segment from o to x we get from

$$\frac{x'(\tau)^2}{(1-x(\tau)^2)^2} \leq \frac{x'(\tau)^2 + y'(\tau)^2}{(1-x(\tau)^2-y(\tau)^2)^2}$$

the inequality

$$L(\gamma_0) \leq L(\gamma).$$

This shows that straight lines through the origin are geodesics. Also

$$\text{(3)} \quad d(o,x) = L(\gamma_0) = \int_o^1 \frac{|x|}{1-t^2x^2}\,dt = \frac{1}{2}\log\frac{1+|x|}{1-|x|}.$$

Consider now the group

$$\mathbf{SU}(1,1) = \{\begin{pmatrix} a & b \\ \bar{b} & \bar{a} \end{pmatrix} : |a|^2-|b|^2 = 1\},$$

which acts on D by means of the maps

$$\text{(4)} \qquad g: z \longrightarrow \frac{az+b}{\bar{b}z+\bar{a}}, \qquad (z \in D).$$

The action is transitive, the subgroup fixing o is $\mathbf{SO}(2)$ so we have the identification

$$\text{(5)} \qquad D = \mathbf{SU}(1,1)\,/\,\mathbf{SO}(2).$$

The Riemannian structure (1) is preserved by the maps (4). In fact, let $z(t)$ be a curve with $z(0) = z$, $z'(0) = u$. Then

$$g\cdot u = \{\frac{d}{dt} g(z(t))\}_{t=0} = \text{the vector } \frac{z'(0)}{(\bar{b}z+\bar{a})^2} \text{ at } g(z)$$

and the desired relation

$$\langle g\cdot u, g\cdot u\rangle_{g(z)} = \langle u,u\rangle_z$$

follows by a simple computation.

The mappings (4) are conformal and map circles (and lines) into circles and lines. It follows that the <u>geodesics in</u> D <u>are the circular arcs perpendicular to the boundary</u> $|z| = 1$.

If z_1, z_2 are any points in D the isometry

$$z \longrightarrow \frac{z-z_1}{1-\bar{z}_1 z}$$

maps z_1 to o and z_2 to the point $(z_2 - z_1) / (1-\bar{z}_1 z_2)$. Following this with a rotation around o, this point is mapped into $|z_2-z_1| / |1-\bar{z}_1 z_2|$ so we deduce from (3)

$$(6) \qquad d(z_1,z_2) = \frac{1}{2}\log \frac{|1-\bar{z}_1 z_2|+|z_2-z_1|}{|1-\bar{z}_1 z_2|-|z_2-z_1|}, \qquad z_1,z_2 \in D.$$

Writing the Riemannian structure (2) in the general form

$$ds^2 = \sum_{i,j} g_{ij}\, dx_i\, dx_j, \qquad g_{ij} = (1-|z|^2)^{-2}\delta_{ij}$$

and putting as usual $\bar{g} = |\det(g_{ij})|$, $g^{ij} = (g_{ij})^{-1}$ (inverse matrix), the <u>Riemannian</u> <u>measure</u>

$$f \longrightarrow \int f\sqrt{\bar{g}}\, dx_1 \ldots dx_n$$

and the <u>Laplace-Beltrami</u> <u>operator</u>

$$L\colon f \longrightarrow \frac{1}{\sqrt{\bar{g}}} \sum_k \partial_k \Big(\sum_i g^{ik} \sqrt{\bar{g}}\, \partial_i f\Big)$$

become, respectively,

(7) $$dz = (1-x^2-y^2)^{-2}\, dx\, dy,$$

(8) $$L = (1-x^2-y^2)^2 \Big(\frac{\partial^2}{\partial x^2} + \frac{\partial^2}{\partial y^2}\Big) .$$

It is a general fact that the Riemannian measure and the Laplace-Beltrami operator are invariant under isometries. For D this can be verified directly since the isometries (4) together with the complex conjugation $z \longrightarrow \bar{z}$ generate all isometries of D. Furthermore, this invariance characterizes the measure dz up to a constant factor and characterizes L in (8) in the sense that each differential operator on D which is $SU(1,1)$-invariant is a polynomial in L.

With these preparations we now consider Problems A, B and C. We shall define some eigenfunctions of L, motivated by the Euclidean case. If $\mu \in \mathbb{C}$ and $\omega \in \mathbf{S}^{n-1}$ the function

(9) $$x \longrightarrow e^{\mu(x,\omega)} \qquad (x \in \mathbb{R}^n)$$

has the following properties:

(i) It is an eigenfunction of the Laplacian $L_{\mathbb{R}^n}$ on $\mathbb{R}^n$.

(ii) It is a <u>plane wave with normal</u> ω, that is, it is constant on each hyperplane perpendicular to ω.

A hyperplane in $\mathbb{R}^n$ is orthogonal to a family of parallel lines. The geometric analog for D is a <u>horocycle</u>, i.e. a circle in D tangential to the boundary $B = \partial D$; in fact such a circle ξ is orthogonal to all geodesics in D tending to the point of contact b. If z is a point on ξ we put

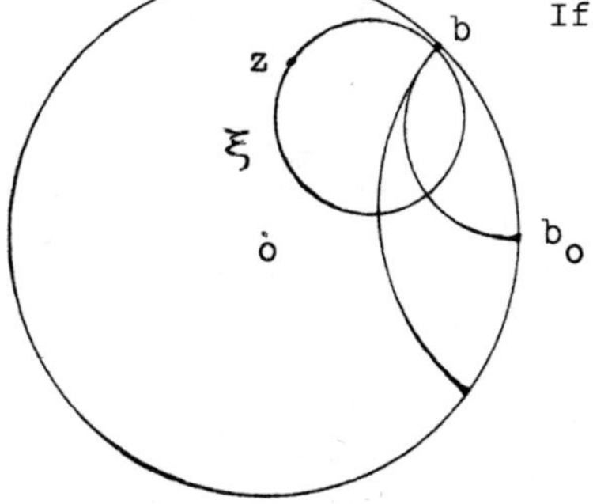

(10) $\langle z,b\rangle$ = distance from o to ξ (with sign) (to be taken negative if o lies inside ξ).

This "inner product" $\langle z,b\rangle$ is in fact the non-Euclidean analog

of (x,ω) which geometrically means the (signed) distance from $0 \in \mathbb{R}^n$ to the hyperplane through x with normal ω. We shall write $\xi(z,b)$ for the above horocycle through z and b.

In analogy with (9) we consider the function

$$e_{\mu,b} : z \longrightarrow e^{\mu<z,b>}, \qquad z \in D. \tag{11}$$

LEMMA 4.1. <u>For the Laplacian</u> L <u>we have</u>

$$L\, e_{\mu,b} = \mu(\mu-2) e_{\mu,b}.$$

<u>Proof</u>. For $t \in \mathbb{R}$ let

$$a_t = \begin{pmatrix} \cosh t & \sinh t \\ \sinh t & \cosh t \end{pmatrix} \in SU(1,1).$$

Then by (3) we find

$$d(o, a_t \cdot o) = d(o, \tanh t) = t. \tag{12}$$

If $b_0 > 0$ is the point of B on the x-axis we therefore have

$$e_{\mu,b_0}(\tanh t) = e^{\mu t}. \tag{13}$$

The horocycles through b_0 are the orbits of the group

$$N = \{n_s = \begin{pmatrix} 1 + is & -is \\ is & 1-is \end{pmatrix} \quad : \quad s \in \mathbb{R}\}.$$

In fact the orbit $N \cdot o$ is such a horocycle and so is the orbit $Na_t \cdot o = a_t N \cdot o$ (a_t normalizes N). Consider now functions on D which, like e_{μ,b_0}, are constant on each of these horocycles. Because of its N-invariance, L maps the class of such functions into itself. The restriction of L to such functions is a differential operator L_0 in the variable t. Since the isometry a_t satisfies $a_t \cdot \tanh \tau = \tanh(t+\tau)$ the invariance of L under a_t means that L_0 is invariant under the translation $\tau \longrightarrow \tau + t$. Thus L_0 has constant coefficients so $e^{\mu t}$ is an eigenfunction for it; hence by (13), e_{μ,b_0} is an eigenfunction of L. The eigenvalue can be calculated by expressing L in the coordinates $a_t n_s \cdot o \longrightarrow (t,s)$; in fact one finds

$$L_0 = \frac{d^2}{dt^2} - 2\,\frac{d}{dt} \,.$$

A simpler computation of the eigenvalue is suggested later.

Reformulating the lemma we have

$$L_z(e^{(i\lambda+1)\langle z,b\rangle}) = -(\lambda^2+1)e^{(i\lambda+1)\langle z,b\rangle}, \qquad \lambda \in \mathbf{C}\,.$$

We now define a Fourier transform on D motivated by the Euclidean case (1) §2. Again dz will denote the surface element (7).

<u>Definition.</u> If f is a complex-valued function on D its

Fourier transform is defined by

$$(14) \qquad \tilde{f}(\lambda,b)=\int_D f(z)e^{(-i\lambda+1)<z,b>}dz$$

for all $\lambda \in \mathbb{C}$, $b \in B$ for which this integral exists.

In order to state the next theorem we call a C^∞ function $\psi(\lambda,b)$ on $\mathbb{C} \times B$, which is holomorphic in λ, a *holomorphic function of uniform exponential type* R *if for each* $N \in \mathbb{Z}^+$

$$\sup_{\lambda \in \mathbb{C},\ b \in B} e^{-R|\mathrm{Im}\lambda|}(1+|\lambda|)^N|\psi(\lambda,b)| < \infty.$$

Here $\mathrm{Im}\,\lambda$ denotes the imaginary part of λ.

We can now state rather explicit answers to Problems A, B and C.

THEOREM 4.2.

(i) *If* $f \in C_c^\infty(D)$ *then*

$$(15) \qquad f(z) = \frac{1}{4\pi}\int_{\mathbb{R}}\int_B \tilde{f}(\lambda,b)e^{(i\lambda+1)<z,b>}\lambda\tanh\left(\frac{\pi\lambda}{2}\right)\,d\lambda\,db,$$

where db *is the circular measure on* B *normalized by* $\int db = 1$.

(ii) *The mapping* $f \longrightarrow \tilde{f}$ *is a bijection of* $C_c^\infty(D)$ *onto the space of holomorphic functions* $\psi(\lambda,b)$ *of uniform exponential type satisfying the functional equation*

$$\int_B e^{(i\lambda+1)<z,b>}\psi(\lambda,b)\,db = \int_B e^{(-i\lambda+1)<z,b>}\psi(-\lambda,b)\,db.$$

(iii) <u>The mapping</u> $f \longrightarrow \tilde{f}$ <u>extends to an isometry of</u> $L^2(D)$ <u>onto</u> $L^2(\mathbb{R}^+ \times B, (2\pi)^{-1}\lambda \tanh(\frac{1}{2}\pi\lambda)d\lambda\, db)$.

<u>Remarks</u>. The three parts of this theorem are non-Euclidean analogs of the inversion formula, the Paley-Wiener theorem, and the Plancherel formula, respectively, for the Fourier transform on $\mathbb{R}^n$. There are however some interesting differences:

(a) While the replacement of the Euclidean measure $\lambda d\lambda d\omega$ in (2) §2 by $\lambda\tanh(\frac{1}{2}\pi\lambda)d\lambda\, db$ is quite natural (cf. (12)) note that in (i) we integrate over $\mathbb{R}$ but in (iii) only over $\mathbb{R}^+$.

(b) While the functional equation in (ii) gives genuine restrictions on the Fourier transform $\psi(\lambda,b) = \tilde{f}(\lambda,b)$ the analogous condition for the Euclidean Fourier transform $\tilde{f}$ would be

$$\int_{S^1} e^{i\lambda(x,\omega)}\tilde{f}(\lambda\omega)d\omega = \int_{S^1} e^{-i\lambda\omega}\tilde{f}(-\lambda\omega)d\omega.$$

This condition is however quite trivial so imposes no restrictions on $\tilde{f}$.

(c) Comparing formulas (14) and (15) with their Euclidean analogs (1), (2) §2 we see the kernel $e^{2<z,b>}$ which does not have any Euclidean analog. If w is the point on the horocycle $\xi(z,b)$ closest to o we have from (3)

$$e^{2<z,b>} = \frac{1+|w|}{1-|w|} .$$

But $|w|$ can be found by using the cosine relation on the triangles (ozb) and (ozc) where c is the center of the horocycle. We obtain

$$\cos(zob) = \frac{1+|z|^2-|z-b|^2}{2|z|} = \frac{|z|^2+(\frac{1}{2}(1+|w|))^2-(\frac{1}{2}(1-|w|))^2}{|z|\,(1+|w|)}$$

so

$$e^{2<z,b>} = \frac{1-|z|^2}{|z-b|^2} , \tag{16}$$

which is just the classical Poisson kernel $P(z,b)$. The Poisson representation formula

$$u(z) = \int_B \frac{1-|z|^2}{|z-b|^2} F(b)\, db$$

for a harmonic function $u(z)$ can therefore be written

$$u(z) = \int_B e^{2<z,b>} F(b)\, db$$

so in accordance with (15) it can be viewed as a formula in non-Euclidean Fourier analysis. We shall see later that this leads to a very simple proof of Schwarz' limit theorem

$$\lim_{u \to b} u(z) = F(b)$$

at each point b of continuity for F. (Theorem 4.20.) It also leads to group-theoretic proof of the classical Fatou theorem that a bounded harmonic function on D has boundary values along almost all radii (cf. Theorem 4.21).

It should be observed that because of (8) the Euclidean and the non-Euclidean harmonic functions coincide. We also note that (16) can be used to give a slightly simpler proof of Lemma 4.1.

In order to describe the solution to Problem B we need the concept of an <u>analytic functional</u> (<u>hyperfunction</u>) on a compact analytic manifold.

Let $\mathcal{A}(B)$ denote the space of analytic functions on the boundary B, considered as an analytic manifold. For each such manifold B, $\mathcal{A}(B)$ carries a certain natural topology; in the case of the circle it can be described rather simply as follows. Let U be an open annulus containing B, $\mathcal{H}(U)$ the space of holomorphic functions on U topologized by uniform convergence on compact subsets. Since each analytic function on B extends to a function in $\mathcal{H}(U)$ for a suitably chosen U, we can identify $\mathcal{A}(B)$ with the union $\bigcup_U \mathcal{H}(U)$ and give it the inductive limit topology. The elements of the dual space $\mathcal{A}'(B)$ are called <u>analytic functionals</u> (or <u>hyperfunctions</u>). Since the elements of $\mathcal{A}'(B)$ generalize measures it is convenient to write

$$T(f) = \int_B f(b)\,dT(b) \qquad f \in A(B), \quad T \in A'(B).$$

For $\lambda \in \mathbb{C}$, let $\mathcal{E}_\lambda(D)$ denote the eigenspace

$$\mathcal{E}_\lambda(D) = \{f \in C^\infty(D) : Lf = -(\lambda^2+1)f\},$$

with the topology induced by that of $C^\infty(D)$.

THEOREM 4.3. *The eigenfunctions of the Laplace-Beltrami operator* L *on* D *are precisely the functions*

$$f(z) = \int_B e^{(i\lambda+1)\langle z,b\rangle}\,dT(b),$$

where $\lambda \in \mathbb{C}$ *and* $T \in A'(B)$. *Moreover, if* $i\lambda \neq -1, -3, -5, \ldots$ *then the mapping* $T \longrightarrow f$ *is a bijection of* $A'(B)$ *onto* $\mathcal{E}_\lambda(D)$.

We shall see later that the hyperfunction T can be recovered from f by a suitable limiting process (at least for $\mathrm{Re}(i\lambda) > 0$) and also that T is a distribution on B if and only if f grows at most exponentially with the distance $d(o,z)$.

For $\lambda \in \mathbb{C}$ let T_λ denote the representation of $SU(1,1)$ on the eigenspace $\mathcal{E}_\lambda(D)$. For Problem C we have the following result.

THEOREM 4.4. *The eigenspace representation* T_λ *is irreducible if and only if*

$$i\lambda + 1 \notin 2\mathbb{Z}.$$

In the remainder of this section we shall give proofs of these theorems.

2. Spherical Functions and Spherical Transforms.

A spherical function on D is by definition a radial eigenfunction of L. If $z \in D$, $r = d(o,z)$ we know from (12) that

$$z = |z|e^{i\theta} = \tanh r \; e^{i\theta}.$$

In the coordinates (r,θ) (geodesic polar coordinates) L has the form

$$(17) \qquad L = \frac{\partial^2}{\partial r^2} + 2 \coth 2r \frac{\partial}{\partial r} + 4 \sinh^{-2}(2r) \frac{\partial^2}{\partial \theta^2} .$$

Thus a spherical function ϕ satisfies

$$(18) \qquad \frac{\partial^2 \phi}{\partial r^2} + 2 \coth 2r \frac{\partial \phi}{\partial r} = -(\lambda^2+1)\phi.$$

Taking for granted the theorem that all the eigenfunctions of the elliptic operator L are analytic we can expand a smooth solution to (18) in a power series in $\sinh(2r)$. It is a special case of Theorem 4.16 proved later that all such solutions are proportional, hence are the constant multiples of the function

(19) $$\phi_\lambda(z) = \int_B e^{(i\lambda+1)\langle z,b\rangle} db.$$

In particular, spherical functions are $\neq 0$ at $z = o$; normalizing spherical functions ϕ such that $\phi(o) = 1$ we see that (19) gives all the spherical functions. Also $\phi_\lambda = \phi_{-\lambda}$ since both functions satisfy (18).

If we substitute $z = \tanh r\, e^{i\phi}, b = e^{i\theta}$ into (16) we obtain from (19)

(20) $$\phi_\lambda(a_r \cdot o) = \frac{1}{2\pi} \int_{-\pi}^{\pi} \left(\cosh 2r - \sinh 2r \cos\theta\right)^{-\frac{1}{2}(i\lambda+1)} d\theta.$$

Since the coefficient $2 \coth 2r$ in (18) has an expansion

$$2 \coth 2r = -2 + 4 \sum_{0}^{\infty} e^{-4nr}$$

it is reasonable to try to find both solutions to (18) by means of such expansions. We consider an expansion

(21) $$\Phi_\lambda(r) = e^{(i\lambda-1)r} \sum_{0}^{\infty} \Gamma_n(\lambda) e^{-2nr}$$

and try to determine the coefficients $\Gamma_n(\lambda)$ such that the function Φ_λ satisfies (18). This leads to the following recursion formula for Γ_n (comparing coefficients to $e^{(i\lambda-1-2n)r}$):

(22) $$n(n-i\lambda)\Gamma_n(\lambda) = \sum_{k\geq 1} \Gamma_{n-2k}(\lambda)(2n-4k-i\lambda+1), \quad (n > 1),$$

where k runs over the integers ≥ 1 for which $n-2k \geq 0$.

Putting $\Gamma_0 \equiv 1$, relation (22) defines Γ_n recursively as a rational function on $\mathbb{C}$. One can also use (22) to estimate the growth of $\Gamma_n(\lambda)$ in n. Let $i\lambda \notin \mathbb{Z}$ and let $h > 0$. Then there exists a constant $K_{\lambda,h}$ such that

$$(23) \qquad |\Gamma_n(\lambda)| \leq K_{\lambda,h} e^{nh} \qquad \text{for all} \quad n \in \mathbb{Z}^+.$$

We shall prove this by induction on the basis of (22). First select constants c_1, c_2 such that for all n

$$|n(n-i\lambda)| \geq c_1 n^2, \ |2n-i\lambda+1| \leq c_2(n+1).$$

Then by (22),

$$|\Gamma_n(\lambda)| \leq \frac{c}{n} \sum_{k \geq 1} |\Gamma_{n-2k}(\lambda)|,$$

where $c = 2 c_2 c_1^{-1}$. Let N_0 be an integer such that

$$c(1-e^{-2h})^{-1} < N_0$$

and select $K = K_{\lambda,h}$ such that

$$|\Gamma_n(\lambda)| \leq K e^{nh} \qquad \text{for} \quad n \leq N_0.$$

Let $N \in \mathbb{Z}^+$, $N > N_0$. Assuming (23) holds for $n < N$ we have

$$|\Gamma_N(\lambda)| \leq \frac{c}{N} \sum_{k\geq 1} K\, e^{(N-2k)h}$$

$$\leq K\, e^{Nh} N^{-1}\, c(1-e^{-2h})^{-1} \leq K\, e^{Nh}$$

and this proves (23) by induction. This estimate shows, h being arbitrary, that the series (21) does converge, and that it gives a solution to (18). But then $\Phi_{-\lambda}$ is another solution and if $i\lambda \notin \mathbb{Z}$ these solutions are linearly independent. It follows that for these λ, ϕ_λ is a linear combination of Φ_λ and $\Phi_{-\lambda}$, say

$$\phi_\lambda = c(\lambda)\Phi_\lambda + c\_(\lambda)\Phi_{-\lambda}.$$

But $\phi_\lambda = \phi_{-\lambda}$ so we conclude $c(\lambda) = c\_(-\lambda)$ so for generic λ

$$\phi_\lambda = c(\lambda)\Phi_\lambda + c(-\lambda)\Phi_{-\lambda}. \tag{24}$$

Going back to the integral representation (20) we can calculate the c-function. Since $\phi_\lambda(a_{-r}\cdot o) = \phi_\lambda(a_r\cdot o) = \phi_{-\lambda}(a_r\cdot o)$ we can replace λ and r in the integral (20) by $-\lambda$ and $-r$. Using the substitution

$$u = \tan \tfrac{1}{2}\theta, \qquad \tfrac{1}{2}d\theta = (1+u^2)^{-1}\, du$$

we obtain

$$(25)\quad \phi_\lambda(a_r\cdot o) = \frac{1}{\pi}\int_{-\infty}^{\infty} (\cosh 2r + \sinh 2r\ \frac{1-u^2}{1+u^2})^{\frac{1}{2}(i\lambda-1)}\ \frac{du}{1+u^2}$$

$$= \frac{1}{\pi}\ e^{(i\lambda-1)r}\int_{-\infty}^{\infty} (1+e^{-4r}u^2)^{\frac{1}{2}(i\lambda-1)}\ (1+u^2)^{-\frac{1}{2}(i\lambda+1)}\ du$$

Assuming $\mathrm{Re}(i\lambda) > 0$ we shall show that the integral tends to a limit as $r \longrightarrow +\infty$. Let $\lambda = \xi + i\eta$ so $\mathrm{Re}(i\lambda) = -\eta > 0$. Select ε $(0 < \varepsilon < \frac{1}{2})$ so small that $1 + 2\varepsilon\eta > 0$. Then the integrand above has the majorization

$$(1+e^{-4r}u^2)^{-\frac{1}{2}\eta-\frac{1}{2}}\ (1+u^2)^{\frac{1}{2}\eta-\frac{1}{2}}$$

$$\leq\ (1+e^{-4r}u^2)^{-\frac{1}{2}\eta+\varepsilon\eta}\ (1+u^2)^{\frac{1}{2}\eta-\frac{1}{2}}$$

$$\leq\ (1+u^2)^{-\frac{1}{2}\eta+\varepsilon\eta}\ (1+u^2)^{\frac{1}{2}\eta-\frac{1}{2}} = (1+u^2)^{-\frac{1}{2}+\varepsilon\eta}$$

and the last expression is integrable. This gives the following result.

THEOREM 4.5. <u>Assume</u> $\mathrm{Re}(i\lambda) > 0$. <u>Then</u>

$$\lim_{r\to+\infty} e^{(-i\lambda+1)r}\ \phi_\lambda(a_r\cdot o) = \pi^{-\frac{1}{2}}\ \frac{\Gamma(\frac{1}{2}i\lambda)}{\Gamma(\frac{1}{2}(i\lambda+1))}\ .$$

In fact, the argument above gives the limit by the dominated convergence theorem as

$$\frac{1}{\pi}\int_{-\infty}^{\infty} (1+u^2)^{-\frac{1}{2}(i\lambda+1)}\ du.$$

Putting $t = (1+u^2)^{-1}$ this becomes

$$\frac{1}{\pi}\int_0^1 t^{\frac{1}{2}(i\lambda+1)} t^{-\frac{1}{2}3}(1-t)^{-\frac{1}{2}}\, dt = \frac{1}{\pi}\frac{\Gamma(\frac{1}{2}i\lambda)\Gamma(\frac{1}{2})}{\Gamma(\frac{1}{2}(i\lambda+1))},$$

and the result follows.

On the other hand, we have from (24)

$$(26)\qquad e^{(-i\lambda+1)r}\phi_\lambda(a_r\cdot o) = c(\lambda)\sum_0^\infty \Gamma_n(\lambda)e^{-2nr} + c(-\lambda)\, e^{-2i\lambda r}\sum_0^\infty \Gamma_n(-\lambda)e^{-2nr}.$$

If $\mathrm{Re}(i\lambda) > 0$ we conclude from (23) and (26), since $\Gamma_0(\lambda) \equiv 1$, that

$$(27)\qquad \lim_{r\to+\infty} e^{(-i\lambda+1)r}\,\phi_\lambda(a_r\cdot o) = c(\lambda).$$

THEOREM 4.6. *The spherical function*

$$\phi_\lambda(z) = \int_B e^{(i\lambda+1)\langle z,b\rangle}\, db \qquad z\in D,\ \lambda\in\mathbb{C}$$

is when $i\lambda \notin \mathbb{Z}$ *given by the following expansion*

$$\phi_\lambda(\tanh r) = c(\lambda)\sum_0^\infty\Gamma_n(\lambda)e^{(i\lambda-1-2n)r} + c(-\lambda)\sum_0^\infty\Gamma_n(-\lambda)e^{(-i\lambda-1-2n)r}$$

where $\Gamma_0 \equiv 1$, Γ_n *is given by* (22) *and*

$$(28) \qquad c(\lambda) = \pi^{-\frac{1}{2}} \frac{\Gamma(\frac{1}{2}i\lambda)}{\Gamma(\frac{1}{2}(i\lambda+1))} \qquad \text{if} \quad -i\lambda \notin \mathbb{Z}^{+}.$$

Only a few remarks are necessary to finish the proof. The assumption that $i\lambda \notin \mathbb{Z}$ guarantees both that all $\Gamma_n(\lambda)$, $\Gamma_n(-\lambda)$ are defined and that Φ_λ and $\Phi_{-\lambda}$ are linearly independent (then the exponentials $e^{(i\lambda-1-2n)r}$ in the first series do not overlap with any exponentials $e^{(-i\lambda-1-2m)r}$ in the second series). Thus the expansion for $\phi_\lambda(\tanh r)$ holds for these λ. On the other hand (24) implies that $c(\lambda)$ and $c(-\lambda)$ are holomorphic for $i\lambda \notin \mathbb{Z}$. In fact given such a λ we can choose suitable r_1, r_2 and solve the linear equations (24) with respect to $c(\lambda)$. On the other hand, (27) shows $c(\lambda)$ holomorphic for $\mathrm{Re}(i\lambda) > 0$. Thus (28) follows.

Definition. If f is a radial function on $\mathbf{D}$ its spherical transform is defined by

$$(29) \qquad \tilde{f}(\lambda) = \int_D f(z)\phi_{-\lambda}(z)\, dz,$$

whenever this integral exists.

Let $I_c^\infty(D)$ denote the set of radial functions in $C_c^\infty(D)$. We also recall that a function F on $\mathbb{C}$ is called an entire function of exponential type R if for each integer $N \geq 0$,

$$(30) \qquad \sup_{\lambda \in \mathbb{C}} e^{-R|\mathrm{Im}\lambda|}(1+|\lambda|)^N |F(\lambda)| < \infty .$$

Let $\mathcal{H}(\mathbb{C})$ denote the set of all such functions F (for varying R).

THEOREM 4.7. *The spherical transform $f \longrightarrow \tilde{f}$ is a bijection of the space $I_c^\infty(D)$ of radial functions of compact support onto the space $\mathcal{H}_e(\mathbb{C})$ of even entire functions of exponential type.*

Along with this theorem of the Paley-Wiener type we shall prove the following inversion and Plancherel formula.

THEOREM 4.8. *The spherical transform $f \longrightarrow \tilde{f}$ is inverted by the formula*

$$(31) \qquad f(z) = \frac{1}{2\pi^2} \int_{\mathbb{R}} \tilde{f}(\lambda)\phi_\lambda(z)|c(\lambda)|^{-2}d\lambda, \quad f \in I_c^\infty(D).$$

Moreover,

$$(32) \qquad \int_D |f(z)|^2\, dz = \frac{1}{2\pi^2} \int_{\mathbb{R}} |\tilde{f}(\lambda)|^2 |c(\lambda)|^{-2} d\lambda.$$

We start by proving a part of Theorem 4.7, then use that part to prove Theorem 4.8 and then go back and finish Theorem 4.7.

Let b_0 be the point of intersection of B with the x-axis. Keeping (12) in mind and letting $d\omega_t$ denote the arc-element on the horocycle ξ_t through $\tanh t$ and b_0 we have by the orthogonality

$$dz = dt\, d\omega_t .$$

It follows from Theorem 4.6 and (29) that

$$(33) \qquad \tilde{f}(\lambda) = \int_{\mathbb{R}} e^{(-i\lambda+1)t}\, dt \int_{\xi_t} f(z)\, d\omega_t(z), \quad f \in I_c^\infty(D).$$

Since $\phi_\lambda(z)$ is even in λ so is $\tilde{f}(\lambda)$. Moreover, the integral over ξ_t has compact support as a function of t so by the classical Paley-Wiener theorem, $\tilde{f} \in \mathcal{H}_e(\mathbb{C})$.

LEMMA 4.9. *Let* F *be an even, entire function of exponential type* R. *Then the function*

$$f(z) = \int_{\mathbb{R}} F(\lambda)\phi_\lambda(z)|c(\lambda)|^{-2}\, d\lambda$$

satisfies

$$f(z) = 0 \quad \textit{if} \quad d(o,z) > R.$$

We write in accordance with Theorem 4.6,

$$\phi_\lambda(\tanh r) = \sum_0^\infty \psi_n(\lambda, r),$$

where

$$\psi_n(\lambda,r) = c(\lambda)\Gamma_n(\lambda)e^{(i\lambda-1-2n)r} + c(-\lambda)\Gamma_n(-\lambda)e^{(-i\lambda-1-2n)r}$$

and shall now prove the following stronger result.

LEMMA 4.10. *Let* F *be as above. Then*

$$\int_{\mathbb{R}} F(\lambda)\psi_n(\lambda,r)|c(\lambda)|^{-2}d\lambda = 0$$

for $r > R$.

Proof. For $\lambda \in \mathbb{R}$ we have

$$|c(\lambda)|^2 = c(\lambda)\overline{c(\lambda)} = c(\lambda)c(-\lambda)$$

so the integral is $e^{-(2n+1)r}$ times

$$\int_{\mathbb{R}} F(\lambda)\Gamma_n(\lambda)c(-\lambda)^{-1}e^{i\lambda r}d\lambda + \int_{\mathbb{R}} F(\lambda)\Gamma_n(-\lambda)c(\lambda)^{-1}e^{-i\lambda r}d\lambda .$$

Since F is even the two integrals actually coincide so let us just consider the last one. If $F(\lambda)\Gamma_n(-\lambda)c(\lambda)^{-1}$ were an entire function of exponential type the lemma would be a consequence of the classical Paley-Wiener theorem. But both $\Gamma_n(-\lambda)$ and $c(\lambda)^{-1}$ have poles. Fortunately there is a half-plane where both are holomorphic. In fact, by the recursion formula for the function $\Gamma_n(\lambda)$ its poles occur for $i\lambda \in \mathbf{Z}^+$; in particular, $\Gamma_n(-\lambda)$ is holomorphic for λ in the lower half plane. By (28) the poles of $c(\lambda)^{-1}$ are the poles of $\Gamma(\frac{1}{2}(i\lambda+1))$ so it is also holomorphic in the lower half plane. We therefore use Cauchy's theorem to shift the contour downwards: For $\eta < 0$

$$(34) \qquad \int_{\mathbb{R}} F(\xi)\Gamma_n(-\xi)c(\xi)^{-1}e^{-i\xi r}\, d\xi$$

$$= \int_{\mathbb{R}} F(\xi+i\eta)\Gamma_n(-\xi-i\eta)c(\xi+i\eta)^{-1}e^{-i\xi r}e^{\eta r}\, d\xi.$$

For this to be permissible we must check the behaviour of the integrand at ∞. First, $\Gamma_n(\lambda)$ is rational. Secondly, $c(\xi+i\eta)^{-1}$ can be estimated by means of known asymtotics for the Gamma function (Magnus-Oberhettinger [1948], p.5) giving for example

$$\left|\frac{\Gamma(z+\frac{1}{2})}{\Gamma(z)}\right| \leq K_1 + K_2|z|^{\frac{1}{2}} \qquad \text{for} \quad \operatorname{Re} z \geq 0.$$

where K_1 and K_2 are constants. This implies by (28)

$$(35) \qquad |c(\xi+i\eta)|^{-1} \leq C_1 + C_2|\xi+i\eta|^{\frac{1}{2}}, \qquad \xi \in \mathbb{R},\ \eta < 0,$$

where C_1 and C_2 are constants. But by (30) the rapid decrease of F cancels out the growth in Γ_n and c at ∞ so the contour-shift above is permissible. Denoting the right hand side of (34) by $Q(r)$ we have

$$|Q(r)| \leq C\, e^{R|\eta|}\, e^{r\eta},$$

where C is a constant (independent of η). Letting $\eta \longrightarrow -\infty$ we deduce $Q(r) = 0$ for $r > R$. This proves the lemma.

In order to prove Lemma 4.9 we substitute the series

$$\phi_\lambda(\tanh r) = \sum_0^\infty \psi_n(\lambda,r)$$

into the integral

$$f(\tanh r) = \int_{\mathbb{R}} F(\lambda)\phi_\lambda(\tanh r)\,|c(\lambda)|^{-2}\,d\lambda.$$

Integrating term-by-term we obtain the desired result $f(\tanh r) = 0$ for $r > R$. The term-by-term integration is justified by the following lemma.

LEMMA 4.11. <u>The Γ-coefficients satisfy</u>

$$|\Gamma_n(\lambda)| \leq c(1+n^d)(1+|\lambda|^e) \qquad \lambda \in \mathbb{R}\ ,n \in \mathbb{Z}^+$$

<u>for suitable constants</u> c,d <u>and</u> e.

<u>Proof</u>. The recursion formula (22) shows that $\Gamma_n(\lambda) \equiv 0$ for n odd. Put

$$a_n(\lambda) = \Gamma_{2n}(\lambda), \quad c_n(\lambda) = |2n(2n-i\lambda)|,$$

$$\gamma_r(\lambda) = |4r-i\lambda+1|, \qquad r \in \mathbb{Z}^+,$$

and define the functions $b_n(\lambda)$ inductively by

$$b_n(\lambda) = c_n(\lambda)^{-1}\sum_{r=0}^{n-1} b_r(\lambda)\gamma_r(\lambda), \qquad b_0(\lambda) \equiv 1.$$

Since

$$2n(2n-i\lambda)a_n(\lambda) = \sum_{r=0}^{n-1} a_r(\lambda)(4r-i\lambda+1),$$

we have by induction

$$|a_n(\lambda)| \leq b_n(\lambda) \qquad \text{for} \quad n \in \mathbb{Z}^+,\ \lambda \in \mathbb{R}.$$

Now

$$b_n(\lambda)c_n(\lambda) = \sum_0^{n-2} b_r(\lambda)\gamma_r(\lambda) + b_{n-1}(\lambda)\gamma_{n-1}(\lambda)$$

$$= b_{n-1}(\lambda)c_{n-1}(\lambda)\left(1+\gamma_{n-1}(\lambda)c_{n-1}(\lambda)^{-1}\right)$$

so by iteration

$$b_n(\lambda)c_n(\lambda) = b_2(\lambda)c_2(\lambda)\prod_{r=2}^{n-1}\left(1+\frac{\gamma_r(\lambda)}{c_r(\lambda)}\right).$$

But it is clear that

$$\gamma_r(\lambda) \leq C\, r^{-1}c_r(\lambda) \qquad r \geq 1,\ \lambda \in \mathbb{R}$$

where C is a constant. Since $\log(1+x) \leq x$ for $x \geq 0$ we conclude that

$$\prod_{r=2}^{n-1}\left(1 + \frac{\gamma_r(\lambda)}{c_r(\lambda)}\right) \leq \exp\left(\sum_2^{n-1} \frac{C}{r}\right) \leq n^C$$

and the lemma follows. Lemma4.9 is now proved as well.

We shall now prove Theorem 4.8 (apart from normalization). Given $f \in C_c(D)$ consider the radial function

$$f^{\natural}(z) = \frac{1}{2\pi}\int_0^{2\pi} f(e^{i\theta}\cdot z)d\theta.$$

Let T denote the linear form

$$\text{(36)} \qquad T: f \longrightarrow \int_{\mathbb{R}} (f^{\natural})^{\sim}(\lambda)|c(\lambda)|^{-2}\, d\lambda, \quad f \in C_c^{\infty}(D).$$

From (33) and (35) it follows that T is a distribution. We shall prove that this distribution T has support at the origin o of D,

$$\text{(37)} \qquad \operatorname{supp}(T) = \{o\}.$$

Let $\beta \in C_c^{\infty}(\mathbb{R})$ have support contained in the interval $-1 \leq r \leq 1$ such that in addition the Fourier transform

$$\text{(38)} \qquad \psi(\lambda) = \int_{\mathbb{R}} \beta(r)e^{-i\lambda r}\, dr$$

satisfies

$$\psi \text{ is even, } \psi \geq 0, \ \psi(0) = 1.$$

Then

$$T(f) = \int_{\mathbb{R}} (f^{\natural})^{\sim}(\lambda)\,|c(\lambda)|^{-2}d\lambda$$

$$= \lim_{\varepsilon \to 0} \int_{\mathbb{R}} (f^{\natural})^{\sim}(\lambda)\psi(\varepsilon\lambda)\,|c(\lambda)|^{-2}\,d\lambda$$

$$= \lim_{\varepsilon \to 0} \int_{\mathbb{R}} \left(\int_{D} f(z)\phi_{-\lambda}(z)dz \right) \psi(\varepsilon\lambda)\,|c(\lambda)|^{-2}\,d\lambda$$

so

(39) $$T(f) = \lim_{\varepsilon \to 0} \int_{D} f(z)T_{\varepsilon}(z)dz,$$

where

(40) $$T_{\varepsilon}(z) = \int_{\mathbb{R}} \psi(\varepsilon\lambda)\phi_{-\lambda}(z)\,|c(\lambda)|^{-2}d\lambda.$$

But the function $\lambda \longrightarrow \psi(\varepsilon\lambda)$ has exponential type ε so by Lemma 4.9 T_{ε} has support inside the closure of the ball $B^{\varepsilon}(o)$ around o with radius ε. Now (39) implies that $\operatorname{supp}(T) = \{o\}$.

Next we prove that T is a measure. For this it suffices to prove that the L^1 norm $\|T_{\varepsilon}\|_1$ is bounded as $\varepsilon \longrightarrow 0$ because then

$$|T(f)| \leq \limsup_{\varepsilon \longrightarrow 0} \int_{D} |f(z)|\;|T_{\varepsilon}(z)|dz$$

$$\leq \limsup_{\varepsilon \longrightarrow 0} \left[\|f|B^{\varepsilon}\|_{\infty} \|T_{\varepsilon}\|_1 \right] \leq \text{const.}\,|f(o)|.$$

Here $|$ denotes restriction and $\|\;\|_{\infty}$ the sup-norm. But $\operatorname{supp}(T_{\varepsilon}) \subset \text{closure}\,(B^{\varepsilon}(o))$ so

$$(41) \qquad \int_D |T_\varepsilon(z)|dz \leq \mathrm{vol}(B^\varepsilon(o))\|T_\varepsilon\|_\infty$$

and

$$\|T_\varepsilon\|_\infty \leq \int_{\mathbb{R}} \psi(\varepsilon\lambda)|c(\lambda)|^{-2}d\lambda = \varepsilon^{-1}\int_{\mathbb{R}} \psi(\lambda)|c(\tfrac{\lambda}{\varepsilon})|^{-2}d\lambda.$$

Thus by (35),

$$\|T_\varepsilon\|_\infty = O(\varepsilon^{-2})$$

and since $\mathrm{vol}(B^\varepsilon(o)) = O(\varepsilon^2)$, (41) implies that $\|T_\varepsilon\|_1$ is bounded as $\varepsilon \longrightarrow 0$.

Being a measure with support $\{o\}$ T is a constant multiple of the delta function at o. Thus

$$(42) \qquad f(o) = c\int_{\mathbb{R}} \tilde{f}(\lambda)|c(\lambda)|^{-2}d\lambda, \qquad f \in I_c^\infty(D),$$

where c is a constant. Given $g \in SU(1,1)$ let us apply (42) to the average

$$F(z) = \int_{\mathbb{K}} f(gk\cdot z)dk, \qquad z \in D,$$

K denoting the group $\mathbf{SO}(2)$. The spherical function ϕ_λ satisfies the functional equation

$$(43) \qquad \int_K \phi_\lambda(gk\cdot z)dk = \phi_\lambda(g\cdot o)\phi_\lambda(z).$$

In fact, as a function of z the left hand side is a radial eigenfunction of L. Hence it is a constant multiple of $\phi_\lambda(z)$. The constant is found by putting $z = o$. Now

$$\tilde{F}(\lambda) = \int_K \left(\int_D f(gk\cdot z)\phi_{-\lambda}(z)dz \right) dk$$

$$= \int_D f(g\cdot z)\phi_{-\lambda}(z)dz = \int_D f(z)\phi_{-\lambda}(g^{-1}\cdot z)dz$$

$$= \int_D f(z) \left(\int_K \phi_{-\lambda}(g^{-1}k\cdot z)dk \right) dz = \phi_{-\lambda}(g^{-1}\cdot o)\tilde{f}(\lambda)$$

so

$$\tilde{F}(\lambda) = \phi_\lambda(g\cdot o)\tilde{f}(\lambda).$$

Since $F(o) = f(g\cdot o)$ we get the desired formula

$$f(g\cdot o) = c \int_{\mathbb{R}} \tilde{f}(\lambda)\phi_\lambda(g\cdot o)|c(\lambda)|^{-2}d\lambda. \tag{44}$$

For the second part of Theorem 4.8 let dg be the Haar measure on $SU(1,1)$ normalized by

$$\int_G f(g\cdot o)dg = \int_D f(z)dz, \qquad f \in C_c^\infty(D).$$

Then if $f_1, f_2 \in I_c^\infty(D)$ the convolution

$$f_1 * f_2(gK) = \int_G f_1(h\cdot o)f_2(h^{-1}g\cdot o)dh$$

has spherical transform

$$(f_1 * f_2)^{\sim}(\lambda) = \int_G f_1(h\cdot o)\left(\int_G f_2(g\cdot o)\phi_{-\lambda}(hg\cdot o)dg\right)dh$$

$$= \int_G f_1(h\cdot o)\left(\int_G f_2(g\cdot o)\left(\int_K \phi_{-\lambda}(hkg\cdot o)dk\right)dg\right)dh$$

$$= \tilde{f}_1(\lambda)\tilde{f}_2(\lambda)$$

because of (43). Applying (42) to the function

$$gK \longrightarrow \int_D f(g\cdot z)\overline{f(z)}dz$$

we derive (32) up to a constant factor which we determine later.

We can now complete the proof of Theorem 4.7. It remains to prove that if $\psi \in \mathcal{H}_e(\mathbf{C})$ and if we define f by

$$f(z) = c\int_{\mathbb{R}} \psi(\lambda)\phi_\lambda(z)|c(\lambda)|^{-2}d\lambda \tag{45}$$

then f (which by Lemma 4.9 is in $I_c^\infty(D)$) has spherical transform ψ. But (44) and (45) imply

$$\int_{\mathbb{R}} (\tilde{f}(\lambda)-\psi(\lambda))\phi_\lambda(z)|c(\lambda)|^{-2}d\lambda \equiv 0.$$

Integration against any $h \in I_c^\infty(D)$ gives

$$\int_{\mathbb{R}} (\tilde{f}(\lambda)-\psi(\lambda))\tilde{h}(\lambda)|c(\lambda)|^{-2}d\lambda = 0.$$

The functions $\tilde{h}$ vanish at ∞ on $\mathbb{R}$. They form an algebra closed under complex conjugation because

$$(h_1 * h_2)^{\sim} = \tilde{h}_1(\lambda)\tilde{h}_2(\lambda)$$

$$(\bar{h})^{\sim}(\lambda) = (\tilde{h}(\lambda))^{-} \qquad (\lambda \in \mathbb{R})$$

because of $\phi_{-\lambda} = \phi_\lambda$. Finally, the algebra separates points on $\mathbb{R}/\mathbb{Z}_2$ because if $\tilde{h}(\lambda_1) = \tilde{h}(\lambda_2)$ for all $h \in I_{\mathbb{C}}^{\infty}(D)$ we have $\phi_{\lambda_1} \equiv \phi_{\lambda_2}$ so applying L we get $\lambda_1^2 + 1 = \lambda_2^2 + 1$. Using Stone-Weierstrass' theorem on $\mathbb{R}/\mathbb{Z}_2$ we get, since $\tilde{f} - \psi$ is even, $\tilde{f} \equiv \psi$. This proves Theorem 4.7.

It remains to determine the normalizing constant in Theorem 4.8. Although this can be done by means of the expansion for ϕ_λ we give instead an entirely different proof of the inversion formula. This proof will also give the normalizing constant.

We note first that the horocycle ξ_t in (33) is the orbit of the point $\tanh t$ under the group

$$N = \{n_x = \begin{pmatrix} 1+ix & -ix \\ ix & 1-ix \end{pmatrix} : x \in \mathbb{R}\}.$$

We find for the image $\zeta = n_x \cdot \tanh t$

$$\zeta = \frac{(1+ix)\tanh t - ix}{ix \tanh t + 1 - ix} = \frac{\operatorname{sh} t - ixe^{-t}}{\operatorname{ch} t - ixe^{-t}}$$

and for the Riemannian measure

$$\cdot \quad dz = \frac{i}{2}\frac{d\zeta \wedge d\bar{\zeta}}{(1-\zeta\bar{\zeta})^2} = e^{-2t}dx \wedge dt.$$

Hence the arc-element dw_t on ξ_t in (33) is $e^{-2t}dx$ so (33) implies

$$\hat{f}(\lambda) = \int_{\mathbb{R}} e^{-i\lambda t}\, e^{-t}\left(\int_{\mathbb{R}} f(\zeta)dx\right)dt.$$

But f is radial so

$$f(\zeta) = f(\tanh s)$$

if

$$(\tanh s)^2 = \zeta\bar{\zeta} = \frac{(\mathrm{sh}t)^2 + x^2e^{-2t}}{(\mathrm{ch}t)^2 + x^2e^{-2t}}.$$

This relation holds if

$$(\mathrm{ch}s)^2 = (\mathrm{ch}t)^2 + x^2e^{-2t}.$$

Defining F by

$$F(\mathrm{ch}s)^2) = f(\tanh s)$$

we have

$$\hat{f}(\lambda) = \int_{\mathbb{R}} e^{-i\lambda t}e^{-t}\left(\int_{\mathbb{R}} F((\mathrm{ch}t)^2 + x^2e^{-2t})dx\right) dt$$

$$= \int_{\mathbb{R}} e^{-i\lambda t}\left(\int_{\mathbb{R}} F((\mathrm{ch}t)^2 + y^2)dy\right) dt.$$

The integral equation

$$\phi(u) = \int_{\mathbb{R}} F(u + y^2)dy$$

(equivalent to Abel's integral equation) can be solved as follows:

$$\int_{\mathbb{R}} \phi'(u+z^2)dz = \int_{\mathbb{R}}\int_{\mathbb{R}} F'(u+y^2+z^2)dy\ dz$$

$$= 2\pi\int_0^{\infty} F'(u+r^2)rdr = \pi \int_0^{\infty} F'(u+\rho)d\rho$$

so

$$-\pi\ F(u) = \int_{\mathbb{R}} \phi'(u+z^2)dz.$$

Thus

$$f(o) = F(1) = -\frac{1}{\pi}\int_{\mathbb{R}} \phi'(1+z^2)dz$$

$$= -\frac{1}{\pi}\int_{\mathbb{R}} \phi'((\operatorname{ch}\tau)^2)\operatorname{ch}\tau\ d\tau.$$

On the other hand we have by the Euclidean Fourier inversion formula

$$\phi((\operatorname{ch} t)^2) = \frac{1}{2\pi}\int_{\mathbb{R}} \hat{f}(\lambda)e^{i\lambda t}d\lambda$$

$$= \frac{1}{2\pi}\int_{\mathbb{R}} \hat{f}(\lambda)\cos\lambda t\ d\lambda$$

since $\hat{f}(\lambda)$ is even. Differentiating with respect to t we obtain

$$-\phi'((\mathrm{cht})^2)2\mathrm{cht}\ \mathrm{sht} = \frac{1}{2\pi}\int_{\mathbb{R}} \hat{f}(\lambda)\lambda\sin\lambda t\ d\lambda.$$

Using the formula

$$\int_{-\infty}^{\infty} \frac{\sin \lambda t}{\mathrm{sht}}\, dt = \pi \tanh\left(\frac{\pi\lambda}{2}\right),$$

which is obtained by integrating the function $z \longrightarrow e^{i\lambda z}/\mathrm{shz}$ over the contour $\mathbb{R} \cup (\mathbb{R} + \pi i)$ we get from the above

$$f(o) = \frac{1}{2\pi^2}\int_{\mathbb{R}} \hat{f}(\lambda)\ \frac{\lambda\pi}{2} \tanh\left(\frac{\pi\lambda}{2}\right)d\lambda.$$

From formulas (28), (49) for the c-function we get the inversion formula (42) with the normalizing constant $(2\pi^2)^{-1}$.

3. The Fourier Transform. Proof of Theorem 4.2.

We shall now prove Theorem 4.2 which includes the inversion formula, the Paley-Wiener theorem and the Plancherel formula for the Fourier transform (14) on the hyperbolic plane D.

LEMMA 4.12. *The convolution*

$$f*\phi_\lambda(z) = \int_G f(g\cdot o)\phi_\lambda(g^{-1}\cdot z)dg$$

<u>is given by</u>

$$f*\phi_\lambda(z) = \int_B e^{(i\lambda+1)\langle z,b\rangle}\tilde{f}(\lambda,b)\,db.$$

<u>Proof</u>. We first observe the geometric identity

$$\langle g\cdot z, g\cdot b\rangle = \langle z,b\rangle + \langle g\cdot o, g\cdot b\rangle \tag{46}$$

for each $g \in SU(1,1)$. For this we consider the horocycles $\xi(g\cdot o, g\cdot b)$ and $\xi(g\cdot z, g\cdot b)$. Using inversion with respect to the point $g\cdot b$ we see that these horocycles cut segments of equal length of the parallel geodesics $(o, g\cdot b)$ and $(g\cdot o, g\cdot b)$. Taking $z = g^{-1}\cdot o$ the identity implies

$$\langle g^{-1}\cdot o, b\rangle = -\langle g\cdot o, g\cdot b\rangle.$$

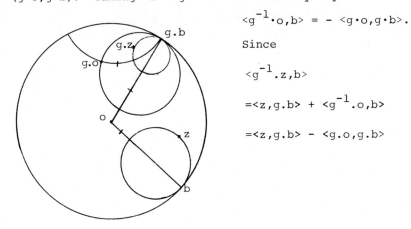

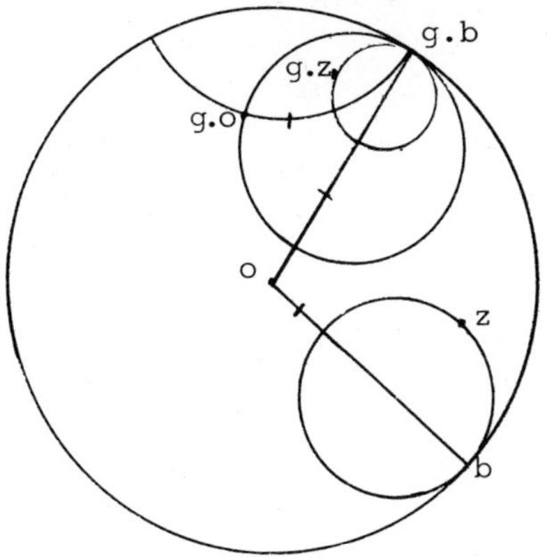

Since

$$\langle g^{-1}.z,b\rangle = \langle z,g.b\rangle + \langle g^{-1}.o,b\rangle = \langle z,g.b\rangle - \langle g.o,g.b\rangle$$

we obtain for the spherical function

$$\phi_\lambda(g^{-1}\cdot z) = \int_B e^{(i\lambda+1)(\langle z,g\cdot b\rangle - \langle g\cdot o,g\cdot b\rangle)}\,db.$$

On the other hand when the conformal mapping g acts on the boundary B the Jacobian is given by

(46a) $$\frac{d(g\cdot b)}{db} = P(g^{-1}\cdot o, b),$$

where P is the Poisson kernel. In fact, if

$$e^{i\phi} = \frac{e^{i\theta}-a}{1-\bar{a}\,e^{i\theta}}$$

a simple computation shows

$$\frac{d\phi}{d\theta} = \frac{1-|a|^2}{|e^{i\theta}-a|^2} \quad .$$

Using the non-Euclidean form of it given by (16) we get by the above

$$\frac{d(g\cdot b)}{db} = e^{2<g^{-1}\cdot o,b>} = e^{-2<g\cdot o,g\cdot b>} \quad .$$

Hence, changing variables in the formula for ϕ_λ, we obtain the symmetric formula

(47) $$\phi_\lambda(g^{-1}\cdot z) = \int_B e^{(i\lambda+1)<z,b>}e^{(-i\lambda+1)<g\cdot o,b>}db.$$

The lemma follows by integrating against $f(g\cdot o)$.

Consider now for a fixed $g \in G$ the radial function

$$f_1(z) = \int_K f(gk\cdot z)dk.$$

From (42) we have

(48) $$f_1(o) = c \int_{\mathbb{R}} \tilde{f}_1(\lambda)\,|c(\lambda)|^{-2}\,d\lambda.$$

But

$$\tilde{f}_1(\lambda) = \int_D \left\{\int_K f(gk\cdot z)\,dk\right\}\phi_{-\lambda}(z)\,dz$$

$$= \int_D f(w)\,\phi_{-\lambda}(g^{-1}\cdot w)\,dw$$

$$= \int_G f(h\cdot o)\,\phi_{-\lambda}(g^{-1}h\cdot o)\,dh = \int_G f(h\cdot o)\,\phi_{\lambda}(h^{-1}g\cdot o)\,dh$$

so

$$\tilde{f}_1(\lambda) = (f*\phi_\lambda)(g\cdot o).$$

Since $f_1(o) = f(g\cdot o)$ we get from Lemma 4.12 and (48) the inversion formula,

$$f(z) = c \int_{\mathbb{R}} \left(\int_B \tilde{f}(\lambda,b)\,e^{(i\lambda+1)\langle z,b\rangle}\,db\right)|c(\lambda)|^{-2}\,d\lambda.$$

But if $\lambda \in \mathbb{R}$ we have by (28)

$$|c(\lambda)|^{-2} = c(\lambda)^{-1}c(-\lambda)^{-1} = \pi\,\frac{\Gamma(\frac{1}{2}(1+i\lambda))\,\Gamma(\frac{1}{2}(1-i\lambda))}{\Gamma(\frac{1}{2}i\lambda)\,\Gamma(-\frac{1}{2}i\lambda)}$$

so

(49) $$|c(\lambda)|^{-2} = \frac{\lambda\pi}{2}\tanh\left(\frac{\pi\lambda}{2}\right)$$

so we have obtained the inversion formula (15).

In order to determine the range of the Fourier transform $f(z) \longrightarrow \tilde{f}(\lambda,b)$ (Theorem 4.2 (ii) and (iii)) we need a definition.

Definition. The point $\lambda \in \mathbb{C}$ is called simple if the mapping $F \longrightarrow f$ of $L^2(B)$ into $C^\infty(D)$ given by

$$(50) \qquad f(z) = \int_B e^{(i\lambda+1)<z,b>} F(b)\,db$$

is one-to-one.

PROPOSITION 4.13. *The point* λ *is non simple if and only if*

$$\lambda = i(1+2k) \qquad k \in \mathbb{Z}^+.$$

Remark. The "conceptual" version of this criterion is that the non simple points λ are the poles of the denominator of the c-function.

We first prove a lemma of Takahashi [1963] .

LEMMA 4.14. *Let* $F(\theta)$ *be integrable on* $0 \leq \theta \leq \pi$ *and let*

$$H(t) = \int_0^\pi (\cosh t - \sinh t \cos\theta)^{-s} F(\theta)\,d\theta$$

If $-s \notin \mathbb{Z}^+$ *then* $H \equiv 0$ *if and only if* $F \equiv 0$.

Proof. We have, by induction on p

$$\frac{d^p}{dt^p}(\cosh t - \sinh t \cos\theta)^{-s}$$

$$= \sum_{q=0}^{p} c_q^p(s)(\cosh t - \sinh t \cos\theta)^{-s-q}(\sinh t - \cosh t \cos\theta)^q$$

where the $c_q^p(s)$ are constants and

$$c_p^p(s) = (-1)^p s(s+1) \cdots (s+p-1).$$

If $-s \notin \mathbb{Z}^+$ then $c_p^p(s) \neq 0$ for all p. Thus if $H \equiv 0$ the equation $H^{(p)}(0) = 0$ implies successively,

$$\int_0^{\pi} F(\theta)\cos^p\theta \, d\theta = 0 \qquad p = 0,1,2,\cdots$$

so $F = 0$ (by the Stone-Weierstrass theorem, say).

Now if $z = \tanh r\, e^{i\phi}$, $b = e^{i\theta}$ formula (50) reads

$$(51) \quad f(z) = \frac{1}{2\pi}\int_0^{2\pi} (\cosh 2r - \sinh 2r \cos\theta)^{-\frac{1}{2}(i\lambda+1)} F(\theta+\phi)\,d\theta.$$

The "if" part of Prof. 4.13 is therefore obvious. On the other hand, suppose

$$-\tfrac{1}{2}(i\lambda+1) \notin \mathbb{Z}^+$$

If $f \equiv 0$ in (51) the proof of Lemma 4.14 implies

$$\int_0^{2\pi} \cos^p \theta \, F(\theta+\phi) \, d\theta = 0, \qquad p \in \mathbb{Z}^+. \tag{52}$$

Expanding F in a Fourier series

$$F(\theta+\phi) \sim \sum a_n e^{in\phi} e^{in\theta} ,$$

(52) implies successively

$$a_p e^{ip\phi} + a_{-p} e^{-ip\phi} = 0 \qquad p = 0,1,2 \cdots$$

so $F = 0$ as desired.

PROPOSITION 4.15. *If* $-\lambda$ *is simple then the function space on* B

$$\{\tilde{f}(\lambda,\cdot) : f \in C_c^\infty(D)\}$$

is dense in $L^2(B)$.

Proof. If this were not so let $F \not\equiv 0$ in $L^2(B)$ satisfy

$$\int_B \tilde{f}(\lambda,b) F(b) \, db = 0 \qquad \text{for all } f \in C_c^\infty(D).$$

This implies

$$\int_B e^{(-i\lambda+1)<z,b>} F(b)\,db = 0, \qquad (z \in D)$$

contradicting the simplicity of $-\lambda$.

Passing now to the proof of Theorem 4.2 (iii) we derive from Lemma 4.12 for λ real

$$(53) \qquad \int_D f*\phi_\lambda(z)\overline{f(z)}\,dz = \int_B |\tilde{f}(\lambda,b)|^2 db.$$

Here we multiply by $|c(\lambda)|^{-2}$, use Lemma 4.12 and the inversion formula and obtain

$$\int_D |f(z)|^2 dz = c \int_{\mathbb{R}\times B} |\tilde{f}(\lambda,b)|^2 |c(\lambda)|^{-2} d\lambda\, db$$

$$= 2c \int_{\mathbb{R}^+\times B} |\tilde{f}(\lambda,b)|^2 |c(\lambda)|^{-2}\, d\lambda\, db,$$

the latter identity coming from evenness in λ which is clear from (53).

Now suppose $F \in L^2(\mathbb{R}^+ \times B, |c(\lambda)|^{-2} d\lambda\, db)$ were orthogonal to the range, i.e.

$$(54) \qquad \int_{\mathbb{R}^+\times B} \tilde{f}(\lambda,b)F(\lambda,b)\ \lambda\tanh(\tfrac{1}{2}\pi\lambda)\,d\lambda\, db = 0, \quad f \in C_c^\infty(D).$$

If $\psi \in I_c^\infty(D)$ we have

$$(55) \qquad (f*\psi)^\sim(\lambda,b) = \tilde{f}(\lambda,b)\tilde{\psi}(\lambda).$$

In fact

$$(f*\psi)^{\sim}(\lambda,b) = \int_D \left\{ \int_G f(g\cdot o)\psi(g^{-1}\cdot z)dg \right\} e^{(-i\lambda+1)<z,b>} dz$$

$$= \int_G f(g\cdot o) \left\{ \int_D \psi(z) e^{(-i\lambda+1)<g\cdot z,b>} dz \right\} dg .$$

Writing now

$$<g\cdot z,b> = <z,g^{-1}\cdot b> + <g\cdot o,b>$$

relation (55) follows.

Replacing now $\tilde{f}(\lambda,b)$ in (54) by $\tilde{f}(\lambda,b)\tilde{\psi}(\lambda)$ we obtain from Theorem 4.7,

$$(56) \qquad \int_B \tilde{f}(\lambda,b)F(\lambda,b)db = 0 \qquad f \in C_c^\infty(D)$$

for all λ outside a null set N_f (depending on f). Thus if $\{f_n\}$ is a countable subset of $C_c^\infty(D)$ equation (56) holds for all $\lambda \in \mathbb{R}^+$ outside a certain fixed null set $N = \bigcup_n N_{f_n}$. We can choose $\{f_n\}$ such that each $f \in C_c(D)$ can be approximated uniformly by a subsequence and such that for each $\lambda \in \mathbb{R}^+$, $\tilde{f}_n(\lambda,b)$ converges to $\tilde{f}(\lambda,b)$ uniformly on B. Thus (56) holds for all $f \in C_c(D)$ and all $\lambda \in \mathbb{R}^+ - N$. But then Prop. 4.15 implies

$$\int_B F(\lambda,b)P(b)db = 0 \qquad \text{for all} \qquad P \in L^2(B)$$

for $\lambda \in \mathbb{R}^+ - N$. Hence $\int_C F(\lambda,b)d\lambda db = 0$ for each "rectangle" C in $(\mathbb{R}^+ - N) \times B$ whence $F = 0$ almost everywhere on $\mathbb{R}^+ \times B$. This proves Part (iii) of Theorem 4.2.

Finally we prove Theorem 4.2 (ii), characterizing the image $C_c^\infty(D)^\sim$. For this we must consider generalizations of the spherical function.

THEOREM 4.16. *Let* $m \in \mathbb{Z}$. *The eigenfunctions* f *of* L *satisfying the homogeneity condition*

$$(57) \qquad f(e^{i\theta}z) = e^{im\theta}f(z)$$

are the constant multiples of the functions

$$\Phi_{\lambda,m}(z) = \int_B e^{(i\lambda+1)\langle z,b\rangle}\chi_m(b)db,$$

where $\lambda \in \mathbb{C}$ *and* $\chi_m(e^{i\phi}) = e^{im\phi}$.

Proof. In view of (17) and (57) the function $F(r) = f(\tanh r)$ satisfies

$$(58) \qquad \frac{d^2F}{dr^2} + 2\coth(2r)\frac{dF}{dr} - 4m^2\sinh^{-2}(2r)F + (\lambda^2+1)F = 0$$

for a suitable λ. Expanding F in a power series of $\sinh(2r)$ we can prove that the smooth solutions to (58) are proportional. In fact, writing

$$F(r) = \sum_0^\infty a_k \sinh^k(2r)$$

we find the recursion formulas

$$m^2 a_0 = 0, \qquad (1-m^2)a_1 = 0$$

$$[(k+2)^2-m^2]a_{k+2} = -[\tfrac{1}{4}(\lambda^2+1) + k(k+1)]a_k, \quad k \geq 0.$$

These equations mean that

$$a_k = 0 \qquad \text{for} \qquad 0 \leq k < |m|;$$

$$a_{|m|} \qquad \text{is arbitrary;}$$

$$a_{|m|+2}, a_{|m|+4}, \ldots \quad \text{are determined by} \quad a_{|m|};$$

$$a_{|m|+1} = a_{|m|+3} = \ldots = 0.$$

This proves the proportionality statement and the proposition follows.

The function $\Phi_{\lambda,m}$ can be expressed in terms of the <u>hypergeometric function</u>. Using (16) and Erdélyi et al. [1953], p.81, we obtain, writing $\nu = \frac{1}{2}(i\lambda+1)$,

$$(59)\ \Phi_{\lambda,m}(|z|) = (1-|z|^2)^{\nu}|z|^{|m|} \frac{\Gamma(|m|+\nu)}{\Gamma(\nu)\,|m|!} F(\nu,|m|+\nu;|m|+1;|z|^2).$$

Using now the transformation formula

$$(59a) \qquad F(a,b;c;z) = (1-z)^{c-a-b}F(c-a,c-b;c;z)$$

we obtain the following invariance property of $\Phi_{\lambda,m}$.

PROPOSITION 4.17. *Let* p_m *denote the polynomial*

$$p_m(x) = \left(|m|-1+\tfrac{1}{2}(x+1)\right) \cdots \left(\tfrac{1}{2}(x+1)\right).$$

Then

$$\Phi_{\lambda,m}(z) = \Phi_{-\lambda,m}(z) \cdot \frac{p_m(i\lambda)}{p_m(-i\lambda)} .$$

We can now obtain a series expansion for $\Phi_{\lambda,m}(\tanh r)$ generalizing Theorem 4.6.

THEOREM 4.18. *The generalized spherical function* $\Phi_{\lambda,m}$ *is when* $i\lambda \notin \mathbb{Z}$ *given by the following expansion*

$$\Phi_{\lambda,m}(\tanh r) = C_1(\lambda)\sum_0^\infty \Gamma_n(\lambda)e^{(i\lambda-1-2n)r} + C_{-1}(\lambda)\sum_0^\infty \Gamma_n(-\lambda)e^{(-i\lambda-1-2n)r},$$

where

$$C_1(\lambda) = c(\lambda), \quad C_{-1}(\lambda) = c(-\lambda)\,\frac{p_m(i\lambda)}{p_m(-i\lambda)} ,$$

$\Gamma_0(\lambda) \equiv 1$ *and* $\Gamma_n(\lambda)$ *is given by the recursion formula*

$$(60) \qquad n(n-i\lambda)\Gamma_n(\lambda) = \sum_{k\geq 1} \Gamma_{n-2k}(\lambda)(2n-4k-i\lambda+1+4m^2k), \quad n \geq 1.$$

Here k *runs over the integers* ≥ 1 *for which* $n-2k \geq 0$.

The proof is almost a repetition of that of Theorem 4.6 so we omit it.

Now let $f \in C_c^\infty(D)$. The functional equation

$$\int_B e^{(i\lambda+1)\langle z,b\rangle} \tilde{f}(\lambda,b)\,db = \int_B e^{(-i\lambda+1)\langle z,b\rangle} \tilde{f}(-\lambda,b)\,db$$

is immediate from Lemma 4.12. Moreover we have in analogy with (33) if $\xi_{t,\theta} = \xi(\tanh t\, e^{i\theta}, e^{i\theta})$

$$\tilde{f}(\lambda, e^{i\theta}) = \int_{\mathbb{R}} e^{(-i\lambda+1)t} \left(\int_{\xi_{t,\theta}} f(z)\,d\omega_t \right) dt,$$

where $d\omega_t$ is the arc-element on the horocycle indicated. This shows that $\tilde{f}$ has uniform exponential type. It just remains to prove that each ψ with the properties in Theorem 4.2 (ii) has the form $\tilde{f}$ for some $f \in C_c^\infty(D)$. For this purpose we define the function $f \in C^\infty(D)$ by

(61) $$f(z) = \frac{1}{2\pi} \int_{\mathbb{R}\times B} \psi(\lambda,b) e^{(i\lambda+1)\langle z,b\rangle} \lambda \tanh(\tfrac{1}{2}\pi\lambda)\,d\lambda\,db.$$

Assuming ψ of uniform exponential type R we shall prove $f(z) = 0$ for $d(o,z) > R$. We consider the expansion

$$\psi(\lambda, e^{i\theta}) = \sum_m \psi_m(\lambda) e^{im\theta}$$

and then

$$f(z) = \sum_m f_m(z),$$

where

(62) $$f_m(z) = \frac{1}{\pi^2} \int_{\mathbb{R}} \Phi_{\lambda,m}(z)\psi_m(\lambda)|c(\lambda)|^{-2}d\lambda.$$

It suffices to prove $f_m(a_r \cdot o) = 0$ for $r > R$. For this we use the expansion of Theorem 4.18.

LEMMA 4.19. _For_ $\varepsilon = \pm 1$ _we have_

$$\int_{\mathbb{R}} C_\varepsilon(\lambda)\Gamma_n(\varepsilon\lambda)e^{\varepsilon i\lambda r}\psi_m(\lambda)|c(\lambda)|^{-2}d\lambda = 0 \quad \text{for} \quad r > R.$$

Proof. First consider the case $\varepsilon = +1$. Since $|c(\lambda)|^2 = c(\lambda)c(-\lambda)$ for $\lambda \in \mathbb{R}$ the integral is

(63) $$\int_{\mathbb{R}} \psi_m(\lambda)\Gamma_n(\lambda)c(-\lambda)^{-1}e^{i\lambda r}d\lambda$$

and this is handled just as the integral in Lemma 4.10 and is found to vanish for $r > R$. For $\varepsilon = -1$ the integral is

(64) $$\int_{\mathbb{R}} \psi_m(\lambda)\Gamma_n(-\lambda)c(\lambda)^{-1}\frac{p_m(i\lambda)}{p_m(-i\lambda)}e^{-i\lambda r}\,d\lambda.$$

Because of the recursion formula (60) and the formula (28) for $c(\lambda)$ we see that $\Gamma_n(-\lambda)c(\lambda)^{-1}$ is holomorphic for λ in the lower half plane. So as in Lemma 4.10 we would like to shift the integration downwards. However, the denominator $p_m(-i\lambda)$ vanishes for $\lambda = -i, -3i, \ldots$ which at first seems to cause a complication. Fortunately $\psi_m(\lambda)$ turns out to have zeros at these points. We shall in fact prove

$$\psi_m(\lambda) = \psi_m(-\lambda)\,\frac{p_m(-i\lambda)}{p_m(i\lambda)} \quad . \tag{65}$$

To see this we integrate the relation

$$e^{(i\lambda+1)<z,b>} = \sum_m \Phi_{\lambda,m}(z)\chi_{-m}(b) \tag{66}$$

against $\psi(\lambda,b)$. This gives

$$\int_B \psi(\lambda,b)e^{(i\lambda+1)<z,b>}db = \sum_m \Phi_{\lambda,m}(z)\psi_m(\lambda)$$

$$= \sum_m \Phi_{\lambda,m}(\tanh r)\ \psi_m(\lambda)e^{im\theta}$$

if $z = \tanh r\, e^{i\theta}$. By our assumption about ψ, the left hand side is even in λ so

$$\Phi_{\lambda,m}(\tanh r)\ \psi_m(\lambda) = \Phi_{-\lambda,m}(\tanh r)\ \psi_m(-\lambda).$$

Now (65) follows from Prop. 4.17. But then the integral (64) reduces to (63) which was already handled so Lemma 4.19 is proved.

Now we substitute the expansion of $\Phi_{\lambda,m}$ into formula (62) for $f_m(a_r.o)$. Since the estimate for $\Gamma_n(\lambda)$ in Lemma 4.11 is still valid we can interchange $\int_{\mathbb{R}}$ and $\sum_n$ and conclude that $f_m(a_r\cdot o) = 0$ for $r>R$. Hence

$$f(z) = 0 \qquad \text{for } d(o,z) > R$$

as desired.

It remains to prove $\tilde{f} = \psi$. By the inversion formula (Theorem 4.2 (i)) we have for $\phi = \tilde{f}-\psi$,

$$\int_{\mathbb{R}} \left(\int_B \phi(\lambda,b) e^{(i\lambda+1)<z,b>} |c(\lambda)|^{-2} db \right) d\lambda \equiv 0.$$

Since the inner integral is even in λ we can here replace $\mathbb{R}$ by $-\mathbb{R}^+$. Integrating the resulting relation against an arbitrary $F \in C_c^\infty(D)$ we obtain

$$\int_{\mathbb{R}^+ \times B} \phi(-\lambda,b)\tilde{F}(\lambda,b)|c(\lambda)|^{-2} d\lambda\, db = 0.$$

But we have seen that the functions $\tilde{F}$ form a dense subset of $L^2(\mathbb{R}^+ \times B)$; hence $\phi \equiv 0$ on $-(\mathbb{R}^+) \times B$. But

$$\int_B \phi(\lambda,b) e^{(i\lambda+1)<z,b>} db = \int_B \phi(-\lambda,b) e^{(-i\lambda+1)<z,b>} db$$

so since each $\lambda \in \mathbb{R}$ simple, $\phi \equiv 0$.

THEOREM 4.20. _Let_ F _be continuous on_ B _and define the harmonic function_ u _on_ D _by the Poisson integral_

$$u(z) = \int_B \frac{1-|z|^2}{|z-b|^2} F(b)\,db.$$

Then

$$\lim_{z \to b_0} u(z) = F(b_0).$$

This classical theorem of H.A. Schwarz is usually proved by reducing it to the case $F(b_0) = 0$; then B is divided up into an ε-arc B_0 around b_0, and its complement. Then u decomposes similarly $u = u_0 + (u-u_0)$. The first term is small near b_0 because of the continutiy of F at b_0, the second term is small since the kernel is small for z near b_0, $|b-b_0| \geq \varepsilon > 0$.

The non-Euclidean point of view gives a simpler proof. Because of (46a) u can be written

$$u(g\cdot o) = \int_B F(g\cdot b)db.$$

It suffices to consider the case $z \longrightarrow b_0$, $b_0 = 1$.

Let a_t be the one-parameter group considered before and k_t rotation by the angle θ_t where $\theta_t \longrightarrow 0$ for $t \longrightarrow +\infty$. Then

$$u(k_t a_t\cdot o) = u(e^{i\theta_t}\tanh t) = \frac{1}{2\pi}\int_0^{2\pi} F(k_t a_t\cdot e^{i\theta})d\theta$$

$$= \frac{1}{2\pi}\int_0^{2\pi} F\left(e^{i\theta_t}\cdot\frac{e^{i\theta}+\tanh t}{\tanh t\, e^{i\theta}+1}\right)d\theta .$$

Letting $t \longrightarrow +\infty$ we get by the dominated convergence theorem

$$\lim_{t\longrightarrow+\infty} u(k_t a_t\cdot o) \longrightarrow \frac{1}{2\pi}\int_0^{2\pi} F(1)d\theta = F(b_o)$$

as desired.

The proof works because the one-parameter group a_t pulls the entire boundary B (except for the point -1) to the point b_0.

We shall now prove the classical theorem of Fatou that a bounded harmonic function on D has radial boundary values almost everywhere on B. While the proof is hardly simpler than the classical one its group theoretic features serve as a basis for the generalization to symmetric spaces.

THEOREM 4.21. (Fatou). _Let_ u _be a bounded harmonic function on_ D. _Then for almost all_ θ

$$\lim_{r \to 1} u(re^{i\theta}) \qquad \underline{\text{exists}}.$$

Proof. It is well known that u is the Poisson integral of a function $F \in L^{\infty}(B)$. Consider now the subgroup Ξ of matrices

$$\Xi = \{\xi_x = \begin{pmatrix} 1+ix & ix \\ -ix & 1-ix \end{pmatrix} : x \in \mathbb{R}\}$$

and the mapping

$$\xi_x \longrightarrow \xi_x \cdot 1 = \frac{1+2ix}{1-2ix}$$

of Ξ into B. It is a bijection of Ξ onto $B - \{-1\}$. Writing $e^{i\theta} = (1+2ix)(1-2ix)^{-1}$ we obtain

$$d\theta = \frac{4dx}{1+4x^2}$$

so

$$\int_B f(b)db = \frac{2}{\pi} \int_{\mathbb{R}} f(\xi_x \cdot 1) \cdot \frac{dx}{1+4x^2} .$$

Now fix $\xi_y \in \Xi$ and use this relation on the function $f(b) = F(\xi_y\, a_t \cdot b)$. Then

$$u(\xi_y\, a_t \cdot o) = \int_B F(\xi_y\, a_t \cdot b)db$$

$$= \frac{2}{\pi} \int_{\mathbb{R}} F(\xi_y\, a_t\, \xi_x \cdot 1) \frac{dx}{1+4x^2}$$

$$= \frac{2}{\pi} \int_{\mathbb{R}} F(\xi_{y+xe^{-2t}} \cdot 1) \frac{dx}{1+4x^2} .$$

Writing $\phi(y) = F(\xi_y)$ we deduce

$$|u(\xi_y\, a_t \cdot o) - F(\xi_y)| \leq \frac{2}{\pi} \int_{\mathbb{R}} |\phi(y+xe^{-2t}) - \phi(y)| \frac{dx}{1+4x^2} .$$

Using now Lebesgue's differentiation theorem

$$\lim_{h \to 0} \frac{1}{2h} \int_{-h}^{h} |\phi(y+x) - \phi(y)|\, dx = 0$$

for almost all y we deduce from the above inequality: for almost all y,

$$\lim_{t \to +\infty} u(\xi_y\, a_t \cdot o) \quad \text{exists.}$$

This means that u has a limit along the geodesic from $\xi_y \cdot o$ to the boundary point $\xi_y \cdot 1$. From this it is easy to deduce that u has the same limit along the radius $(o, \xi_y \cdot 1)$. For this we use the elementary inequality

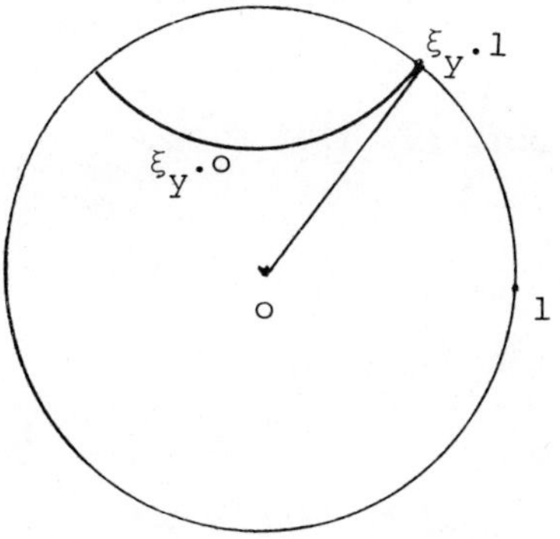

$$|(\nabla u)(z)| \leq \frac{C}{R} \quad \text{for } |z| \leq 1-R$$

∇u being the gradient of u and C being a constant. (This estimate follows immediately from the Poisson formula for the disk $|\zeta - z| \leq R$.) But as R tends to 0 the two geodesics above come together at a faster rate so the inequality implies that u has the same limit along both geodesics. This concludes the proof.

4. Eigenfunctions and Eigenspace Representations. Proofs of Theorems 4.3-4.4.

Given $\lambda \in \mathbb{C}$ we consider now the representation T_λ of $G = SU(1,1)$ on the eigenspace

$$\mathcal{E}_\lambda(D) = \{f \in C^\infty(D) : Lf = -(\lambda^2+1)f\}.$$

If $g \in G$, $T_\lambda(g)$ is the linear transformation of $\mathcal{E}_\lambda(D)$ given by

$$(T_\lambda(g)f)(z) = f(g^{-1}\cdot z).$$

We shall now prove Theorem 4.4 that

(67) $\quad T_\lambda$ is irreducible if and only if $i\lambda+1 \notin 2\mathbb{Z}$.

Remark. T_{λ_o} is irreducible if and only if λ_o is not a pole of the denominator of the function $c(\lambda)c(-\lambda)$.

LEMMA 4.22. *The point* $\lambda \in \mathbb{C}$ *is simple if and only if the functions*

$$b \longrightarrow \sum_k a_k e^{(i\lambda+1)\langle z_k, b\rangle} \qquad a_k \in \mathbb{C},\ z_k \in D$$

form a dense subspace of $L^2(B)$.

This is an immediate reformulation of the definition of simplicity.

LEMMA 4.23. *The mapping* $F \longrightarrow f$ *of* $L^2(B)$ *into* $\mathcal{E}_\lambda(D)$ *given by*

$$f(z) = \int_B e^{(i\lambda+1)\langle z,b\rangle} F(b)\,db \tag{68}$$

is continuous.

The topology of $C^\infty(D)$ is described in §1. The lemma follows immediately by using Schwarz' inequality.

For $\lambda \in \mathbb{C}$ simple let $\mathcal{H}_\lambda$ denote the space of functions f defined by (68) as F runs through $L^2(B)$. Then $\mathcal{H}_\lambda$ is a Hilbert space if f is given the L^2 norm of F.

LEMMA 4.24. <u>Suppose</u> $\lambda \in \mathbb{C}$ <u>is simple</u>. <u>Then</u> $\mathcal{H}_\lambda$ <u>is dense in</u> $\mathcal{E}_\lambda(D)$.

The proof is analogous to that of Lemma 2.8 so we omit it.

Suppose now $i\lambda+1 \notin 2\mathbb{Z}$. This means that both λ and $-\lambda$ are simple. Let $0 \neq V \subset \mathcal{E}_\lambda(D)$ be a closed invariant subspace. Then V contains an element f such that $f(o) \neq 0$; averaging over the rotations around o we conclude $\phi_\lambda \in V$. Now by (47),

$$\sum_k a_k \phi_\lambda(g_k^{-1} \cdot z) = \int_B e^{(i\lambda+1)\langle z,b\rangle} \sum_k a_k e^{(-i\lambda+1)\langle g_k \cdot o,b\rangle}\, db. \tag{69}$$

Since $-\lambda$ is simple we conclude from Lemma 4.22-4.23 that the closed subspace of $\mathcal{E}_\lambda(D)$ generated by the various linear combinations (69) contains $\mathcal{H}_\lambda$. But then Lemma 4.24 shows $V = \mathcal{E}_\lambda(D)$. Hence T_λ is irreducible.

On the other hand, suppose T_λ is irreducible. Since $T_\lambda = T_{-\lambda}$, since the condition $i\lambda+1 \notin 2\mathbb{Z}$ is invariant under $\lambda \longrightarrow -\lambda$, and since λ or $-\lambda$ is simple we may assume λ simple.

Let $0 \neq E \subset \mathcal{H}_\lambda$ be a closed invariant subspace of the Hilbert space $\mathcal{H}_\lambda$. By the irreducibility of T_λ, E is dense in $\mathcal{E}_\lambda(D)$. Let

$$f*\chi_m(z) = \frac{1}{2\pi}\int_0^{2\pi} f(e^{-i\theta}\cdot z)\chi_m(e^{i\theta})\,d\theta.$$

Then E being closed in $\mathcal{H}_\lambda$ and invariant,

$$(70) \qquad E*\chi_m \subset E.$$

By continuity of the map $f \longrightarrow f*\chi_m$,

$$(71) \qquad E*\chi_m \text{ is dense in } \mathcal{E}_\lambda(D)*\chi_m.$$

But by Theorem 4.16 this last space equals $\mathbb{C}\Phi_{\lambda,m}$. Thus (70)-(71) imply $\Phi_{\lambda,m} \in E$ whence $E = \mathcal{H}_\lambda$. Thus G acts irreducibly on $\mathcal{H}_\lambda$ so the functions (69) are dense in $\mathcal{H}_\lambda$. Using Lemma 4.22 we conclude that $-\lambda$ is simple. This concludes the proof of Theorem 4.4.

Next we prove Theorem 4.3 giving the integral representation of the eigenfunctions of L by means of analytic functionals. Let f be any eigenfunction of L and select λ simple such that

$$Lf = -(\lambda^2+1)f.$$

For fixed $z \in D$ we develop the function $\theta \longrightarrow f(e^{i\theta}z)$ in an absolutely convergent Fourier series

(72) $$f(e^{i\theta}z) = \sum_n c_n(z)e^{in\theta},$$

where

$$c_n(z) = \frac{1}{2\pi}\int_0^{2\pi} f(e^{i\theta}z)e^{-in\theta}d\theta.$$

Then by Theorem 4.16,

$$c_n(z) = a_n\Phi_{\lambda,n}(z), \quad a_n \in \mathbb{C}.$$

Using the transformation formula

$$F(a,b;c;z) = (1-z)^{-a}F(a,c-b;c;\frac{z}{z-1})$$

for the hypergeometric function (Erdélyi et al. [1953], p.64) we derive from (59) with $\nu = \frac{1}{2}(i\lambda+1)$,

(73) $$\Phi_{\lambda,n}(r) = r^{|n|}\frac{\Gamma(|n|+\nu)}{\Gamma(\nu)|n|!}F(\nu,1-\nu;|n|+1;\frac{r^2}{r^2-1}),$$

where $r = |z|$. We know from (72) that

(74) $$\sum_n |a_n\Phi_{\lambda,n}(r)| < \infty$$

and we want to deduce from this that

(75) $$\sum_n |a_n|r^{|n|} < \infty, \quad 0 < r < 1.$$

Let us choose $k \in \mathbb{Z}^+$ such that $k > |\nu|$ and put

$x = r^2(r^2-1)^{-1}$. Then for $n > 0$

$$F(\nu,1-\nu;n+1;x) = p_k(n,x) + \rho_{k+1}(n,x),$$

where p_k is the kth Taylor polynomial,

$$p_k(n,x) = 1+ \frac{\nu(1-\nu)}{(n+1)} x + \cdots \frac{\nu(\nu+1)\cdots(\nu+k-1)(1-\nu)\cdots(1-\nu+k-1)}{(n+1)\cdots(n+k)k!} x^k$$

and ρ_{k+1} the remainder. Using the general formula

$$R_k(x) = \frac{1}{k!}\int_0^1 x^{k+1}(1-s)^k F^{(k+1)}(xs)ds$$

for the remainder term in a Taylor series, and the Euler integral formula for the hypergeometric function

$$F(a,b;c;z) = \frac{\Gamma(c)}{\Gamma(b)\,\Gamma(c-b)}\int_0^1 t^{b-1}(1-t)^{c-b-1}(1-tz)^{-a}dt, \quad (\mathrm{Re}\,c>\mathrm{Re}\,b>0)$$

we obtain for $n+1 > \mathrm{Re}(1-\nu) > 0$

$$\rho_{k+1}(n,x) =$$

$$\frac{n!\,\Gamma(k+\nu+1)x^{k+1}}{\Gamma(1-\nu)\Gamma(n+\nu)\Gamma(\nu)k!}\int_0^1\int_0^1 t^{1-\nu+k}(1-t)^{n+\nu-1}(1-s)^k(1-stx)^{-\nu-k-1}ds\,dt\,.$$

By holomorphic continuation in ν this will hold for $n > |\nu|$. Since $x < 0$, $0 \le s,t \le 1$, and $k > |\nu|$ we have

$$|(1-st\,x)^{-\nu-k-1}| \le 1,$$

so

$$|\rho_{k+1}(n,x)| \leq |x|^{k+1} \frac{n!\ \Gamma(k+\nu+1)}{|\Gamma(1-\nu)\Gamma(n+\nu)\Gamma(\nu)|k!} \frac{\Gamma(2-\mathrm{Re}\nu+k)\Gamma(n+\mathrm{Re}\nu)}{\Gamma(2+n+k)}.$$

Using the asymptotic property

$$\lim_{|z|\longrightarrow\infty} e^{-\alpha \log z} \frac{\Gamma(z+\alpha)}{\Gamma(z)} = 1$$

of the Gammafunction we have

$$\lim_{n\longrightarrow\infty} \frac{\Gamma(n+\mathrm{Re}\,\nu)}{|\Gamma(n+\nu)|} = 1$$

and consequently,

$$(76) \qquad |\rho_{k+1}(n,x)| \leq C_k |x|^{k+1}\, n^{-(k+1)},$$

where C_k is a constant. Now we put

$$b_n = |a_n| \left|\frac{\Gamma(n+\nu)}{\Gamma(\nu)n!}\right|, \quad \varepsilon(n,x) = \frac{\rho_{k+1}(n,x)}{C_k|x|^{k+1}} n^{k+1}.$$

Then by (74)

$$\sum_{n=0}^{\infty} b_n r^n |p_k(n,x)+\rho_{k+1}(n,x)| < \infty ,$$

whence

$$(77) \qquad \sum_{n=0}^{\infty} b_n r^n\, n^{-(k+1)} \left|\frac{n^{k+1}}{|x|^{k+1}C_k} p_k(n,x)+ \varepsilon(n,x)\right| < \infty.$$

We have fixed ν and k. Let $0 < r_0 < 1$ be arbitrary and put $x_0 = r_0^2/(r_0^2 - 1)$. Then select N_{r_0} such that

$$\left| \frac{n^{k+1}}{|z_0|^{k+1} C_k} p_k(n,z_0) \right| \geq 2 \qquad \text{for} \quad n \geq N_{r_0}.$$

This can be done since

$$\lim_{n \to \infty} p_k(n,z_0) = 1.$$

But $|\varepsilon(n,x)| \leq 1$ so by (77)

$$\sum_0^\infty |b_n r_0^n n^{-(k+1)}| < \infty.$$

Since r_0 was arbitrary this proves

$$\sum_0^\infty |a_n| r^n < \infty \qquad 0 \leq r < 1.$$

But this proves (75) since n appears in (73) only as $|n|$.

Now given a hyperfunction $T \in A'(B)$ we define its Fourier series by

$$T \sim \sum_n \alpha_n e^{in\theta}, \quad \text{if} \quad \alpha_n = \int_0^{2\pi} e^{-in\theta} dT(\theta).$$

Then we have the following result.

LEMMA 4.25. <u>A hyperfunction is uniquely determined by its Fourier series. The series</u>

$$\sum_n \alpha_n e^{in\theta}$$

<u>is the Fourier series of a hyperfunction</u> $T \in A'(B)$ <u>if and only if</u>

(78) $$\sum_n |\alpha_n| r^{|n|} < \infty \quad \underline{\text{for all}} \quad 0 \leq r < 1.$$

<u>Proof</u>. Suppose we have a Laurent series

(79) $$F(z) = \sum_n b_n z^n$$

converging in an annulus containing $B\colon |z| = 1$. This is equivalent to both series

$$\sum_0^\infty b_n z^n, \quad \sum_0^\infty b_{-n} z^n$$

having radius of convergence > 1 or equivalently,

(80) $$\limsup_n |b_n|^{\frac{1}{n}} = \beta < 1, \ \limsup_n |b_{-n}|^{\frac{1}{n}} = \gamma < 1;$$

all such numbers β, γ $0 \leq \beta, \gamma < 1$ can occur.

The series (79) converges in the topology of $A(B)$. Thus if $T \in A'(B)$ has Fourier coefficients (α_n) we have

$$T(F) = \sum_n b_n \alpha_{-n}.$$

In particular, T is determined by its Fourier series. Because of (80) and the arbitrariness of β and γ we deduce

$$\sum_n |\alpha_n| r^{|n|} < \infty \qquad 0 \le r < 1.$$

On the other hand if this condition is satisfied we can define a linear form T on $A(B)$ by

$$T(F) = \sum_n \alpha_n \, b_{-n}.$$

To see that T is continuous we represent $A(B)$ as the union

$$A(B) = \bigcup_{n=1}^{\infty} \mathcal{H}_n,$$

where $\mathcal{H}_n$ is the set of holomorphic functions in the annulus $1 - \frac{1}{n} < |z| < 1 + \frac{1}{n}$. By the definition of the topology of $A(B)$ we must prove that for each n, the restriction $T|\mathcal{H}_n$ is continuous (for the topology of uniform convergence on compact sets). Because of (78) the functions

$$f_1(z) = \sum_0^{\infty} \alpha_k \, z^k, \quad f_2(z) = \sum_1^{\infty} \alpha_{-k} \, z^k$$

are holomorphic in the unit disk $|z| < 1$. For the curves

$$C_1: \ |z| = 1 - \frac{1}{2n}, \quad C_2: \ |z| = 1 + \frac{1}{2n}$$

we have

$$\sum_{0}^{\infty} \alpha_k \, b_{-k} = \frac{1}{2\pi i} \int_{C_1} F(\zeta) f_1(\zeta) \frac{d\zeta}{\zeta} \, , \quad F \in \mathcal{H}_n$$

$$\sum_{1}^{\infty} \alpha_{-k} \, b_k = \frac{1}{2\pi} \int_{C_2} F(\zeta) f_2(\tfrac{1}{\zeta}) \frac{d\zeta}{\zeta} \, , \quad F \in \mathcal{H}_n$$

from which the continuity of $T|\mathcal{H}_n$ is obvious. This proves the lemma.

Going back to the proof of Theorem 4.3 we know now that there exists a hyperfunction T with Fourier series

$$T \sim \sum_n a_n \, e^{in\theta} \, .$$

But given $z \in D$ the function

$$\zeta \longrightarrow \left(\frac{1 - z\,\bar{z}}{(\zeta - z)(\zeta^{-1} - \bar{z})} \right)^{\nu} \qquad \nu = \tfrac{1}{2}(i\lambda + 1)$$

is holomorphic in an annulus U containing B and can be expanded there in a Laurent series. The restriction of this series to B is the Fourier series (66). Since the Laurent series converges uniformly on compact subsets and since the injection $\mathcal{H}(U) \longrightarrow A(B)$ is continuous it follows that (66) converges in the topology of $A(B)$. Hence we can apply T to it term-by-term and deduce from (72)

$$\int_B e^{(i\lambda+1)\langle z,b\rangle} dT(b) = \sum_m \Phi_{\lambda,m}(z) a_m = f(z),$$

which is the desired integral representation of f.

On the other hand we must verify that if $T \in A'(B)$

the function

$$f(z) = \int_B e^{(i\lambda+1)\langle z,b\rangle} dT(b) \tag{81}$$

is an eigenfunction of L. The continuity of f is clear since $z_n \longrightarrow z_0$ implies $e^{(i\lambda+1)\langle z_n,b\rangle} \longrightarrow e^{(i\lambda+1)\langle z_0,b\rangle}$ in the topology of $A(B)$. Next we use the following result whose proof is entirely analogous to that of Prop. 2.5. Let dk be the normalized Haar measure on $K = \mathbf{SO}(2)$.

PROPOSITION 4.26. *A function f satisfying $Lf = -(\lambda^2+1)f$ satisfies the functional equation*

$$\int_K f(gk\cdot z)dk = f(g\cdot o)\phi_\lambda(z), \qquad g \in G, \ z \in D.$$

Conversely, a continuous function f satisfying this functional equation is automatically of class C^∞ and satisfies $Lf = -(\lambda^2+1)f$.

That the function (81) satisfies this functional equation is verified in the same way as its continuity. This then concludes the proof of Theorem 4.3.

Remark. If λ is simple then the mapping $T \in A'(B) \longrightarrow f \in \mathcal{E}_\lambda(D)$ is one-to-one.

In fact, f and T are related by

(82) $$f(z) = \sum_n a_n \Phi_{\lambda,n}(z), \quad T \sim \sum a_n e^{in\theta},$$

and

$$a_n \Phi_{\lambda,n}(z) = \frac{1}{2\pi}\int_0^{2\pi} f(e^{i\theta})e^{-in\theta}\, d\theta.$$

By the simplicity of λ each $\Phi_{\lambda,n} \not\equiv 0$ so f determines the sequence (a_n) which by Lemma 4.25 determines T.

We shall now prove that for $\mathrm{Re}(i\lambda) > 0$ the hyperfunction T can be recovered from f by a suitable limit process, generalizing (27).

THEOREM 4.27. _Assuming_ $\mathrm{Re}(i\lambda) > 0$ _we have_

$$\lim_{d(0,2) \to \infty} e^{(1-i\lambda)d(o,z)} f(z) = c(\lambda)T$$

in the sense that for each n

(83) $$\lim_{t \to +\infty} e^{(1-i\lambda)t} a_n \Phi_{\lambda,n}(e^{i\theta}\tanh t) = c(\lambda)a_n e^{in\theta} .$$

Proof. From (59) and (59a) we have

(84) $$\Phi_{\lambda,n}(|z|) =$$

$$(1-|z|^2)^{1-\nu} |z|^{|n|} \frac{\Gamma(|n|+\nu)}{\Gamma(\nu)|n|!} F(|n|+1-\nu, 1-\nu; |n|+1; |z|^2).$$

For $\mathrm{Re}\, c > \mathrm{Re}(a+b)$, $-c \notin \mathbb{Z}^+$, the power series (around $z = 0$) for the hypergeometric function $F(a,b;c;z)$ converges absolutely for $|z| \leq 1$ and

$$(85) \qquad F(a,b;c;1) = \frac{\Gamma(c)\Gamma(c-z-b)}{\Gamma(c-a)\Gamma(c-b)} .$$

The conditions are satisfied for the hypergeometric function in (84) since

$$c = |n|+1 > 0, \quad \mathrm{Re}\left\{|n|+1-(|n|+1-\nu)-(1-\nu)\right\} = \mathrm{Re}(i\lambda) > 0.$$

Hence, putting $z = \tanh t$ we obtain

$$(86) \quad \lim_{t \to \infty} (\mathrm{ch}t)^{2-2\nu} \Phi_{\lambda,n}(\tanh t) = \frac{\Gamma(|n|+\nu)}{\Gamma(\nu)\,|n|!} \cdot \frac{\Gamma(|n|+1)\Gamma(2\nu-1)}{\Gamma(\nu)\Gamma(|n|+\nu)}$$

$$= \frac{\Gamma(2\nu-1)}{\Gamma(\nu)^2} .$$

Using the duplication formula for the Γ function and the formula (28) for $c(\lambda)$, formula (83) follows.

Under a mild condition on λ we shall now prove that the eigenfunction $f(z)$ grows at most exponentially with the distance $d(o,z)$ if and only if the hyperfunction T is a distribution. We shall need the following analog of Lemma 4.25.

LEMMA 4.28. *The series*

$$\sum_n \alpha_n e^{in\theta}$$

is the Fourier series of a distribution T *on* B *if and only if*

$$|\alpha_n| \leq C(1+|n|)^{\ell} \qquad (n \in \mathbb{Z})$$

for some constants C _and_ ℓ.

The proof (cf. Schwartz [1966], Ch. VII) is an easy consequence of the fact that every distribution T on B is a finite sum of derivatives (in the sense of distributions) of continuous functions f_k. The Fourier series for T is obtained by differentiating the Fourier series for the f_k so the estimate for α_n follows. Conversely if the series $\sum \alpha_n e^{in\theta}$ has α_n satisfying the above estimates the series

$$\sum_n \alpha_n (1+n^2)^{-k} e^{in\theta}$$

is absolutely convergent (for $k \in \mathbb{Z}^+$ large enough) to a continuous function f. The given series then represents a suitable distribution derivative of f.

Given $\lambda \in \mathbb{C}$ let $\mathcal{E}^*_\lambda(D)$ denote the subspace of functions $f \in \mathcal{E}_\lambda(D)$ satisfying an inequality

$$(87) \qquad |f(z)| \leq C\, e^{ad(o,z)}, \qquad z \in D,$$

for some constants C and a.

THEOREM 4.29. Assume $i\lambda \notin \mathbf{Z}$. Then the mapping $T \longrightarrow f$, where

$$f(z) = \int_B e^{(i\lambda+1)<z,b>} dT(b),$$

is a bijection of the space $\mathcal{D}'(B)$ of distributions on B onto $\mathcal{E}^*_\lambda(D)$.

Proof. Consider the Fourier expansion (66). If L_B is the Laplacian $d^2/d\theta^2$ on B we have

$$(-L_B)^p_b \left(e^{(i\lambda+1)<z,b>}\right) = \sum_n \Phi_{\lambda,n}(z)\, n^{2p}\, \chi_{-n}(b)$$

so

$$(88) \qquad \Phi_{\lambda,n}(z) n^{2p} = \int_B (-L_B)^p_b \left(e^{(i\lambda+1)<z,b>}\right) \chi_n(b) db.$$

If $z = \tanh t\; e^{i\phi}$, $b = e^{i\theta}$ we have by (16)

$$(89) \qquad e^{2\nu<z,b>} = [ch(2t) - sh(2t)\cos(\phi-\theta)]^{-\nu}.$$

From the definition of $<z,b>$ it is easily seen that

$$|<z,b>| \leq d(o,z) \quad \text{for all} \quad z,b.$$

Thus if we apply L_B^p to (89) it follows quickly that

$$|L_B^p(e^{2\nu<z,b>})| \leq C\, e^{ad(o,z)},$$

where C and a are constants depending on p and ν. Thus by (88)

$$|\Phi_{\lambda,n}(z)n^{2p}| \leq C\, e^{ad(o,z)}. \tag{90}$$

Suppose now $T \in \mathcal{D}'(B)$ has Fourier coefficients a_n. Since (66) converges in the topology of $\mathcal{E}(B)$ we can apply T to it term-by-term and deduce

$$f(z) = \sum_n \Phi_{\lambda,n}(z)a_n. \tag{91}$$

But by Lemma 4.28 a_n is bounded by a power of n and since (90) holds for each $p \in \mathbb{Z}^+$ it follows that $f \in \mathcal{E}^*_\lambda(D)$.

Conversely, suppose λ satisfies the conditions of the theorem and suppose $f \in \mathcal{E}^*_\lambda(D)$. Let $T \in A'(B)$ represent f in the sense that

$$f(z) = \int_B e^{(i\lambda+1)\langle z,b\rangle}dT(b).$$

Since λ is simple we know by a previous remark that T is unique.

In order to prove that the a_n satisfy the condition of Lemma 4.28 we shall establish a modified version of (83). For this we need the following relation

$$F(a,b;c;x) = \frac{\Gamma(c)\Gamma(c-a-b)}{\Gamma(c-a)\Gamma(c-b)} F(a,b;a+b-c+1;1-x)$$

$$+ (1-x)^{c-a-b} \frac{\Gamma(a+b-c)\Gamma(c)}{\Gamma(a)\Gamma(b)} F(c-a,c-b;c-a-b+1;1-x)$$

(cf. Erdélyi et al. [1953], p.108) valid for $-c \notin \mathbb{Z}^+$, $a+b-c \notin \mathbb{Z}$. Using this on (59) we obtain after simple manipulations (with $\nu = \frac{1}{2}(i\lambda+1)$),

$$\Phi_{\lambda,m}(\tanh t) =$$

$$\tanh^{|m|}t \; ch^{-2\nu}t \; \frac{\Gamma(1-2\nu)\Gamma(|m|+\nu)}{\Gamma(\nu)\Gamma(1-\nu)\Gamma(|m|+1-\nu)} \; F(|m|+\nu,\nu;2\nu;ch^{-2}t)$$

$$+ \tanh^{|m|}t \; ch^{2\nu-2}t \; \frac{\Gamma(2\nu-1)}{\Gamma(\nu)^2} \; F(|m|-\nu+1,1-\nu;2-2\nu;ch^{-2}t)$$

valid if $2\nu \notin \mathbb{Z}$ (which is equivalent to our assumption about λ).

Moreover, if ${}_1F_1$ denotes, as usual, the <u>confluent hypergeometric function</u> we have for each $|x| < 1$

$$\lim_{m \to +\infty} F(a+m,b;c;x/m) = {}_1F_1(b,c;x) \tag{92}$$

(Erdélyi et al. [1953], p.248). We construct now the sequence $t_m \longrightarrow \infty$ by

$$ch^2 t_m = |m|/a \,, \tag{93}$$

where $a > 0$ is a number to be determined later. Using (92) and the relation

$$\lim_{m \to +\infty} \frac{\Gamma(x+m)}{\Gamma(m)} m^{-x} = 1,$$

we deduce from the formula for $\Phi_{\lambda,m}$,

$$\lim_{|m| \to \infty} (\mathrm{ch}\, t_m)^{2-2\nu} \Phi_{\lambda,m}(\tanh t_m)$$

$$= e^{-\frac{1}{2}a} \frac{\Gamma(1-2\nu)}{\Gamma(\nu)} \left[\frac{a^{2\nu-1}}{\Gamma(1-\nu)} \, {}_1F_1(\nu,2\nu;a) + \frac{1}{\Gamma(\nu)} \, {}_1F_1(1-\nu,2-2\nu;a) \right] .$$

Since

$$\lim_{x \to +\infty} {}_1F_1(b,c,x) e^{-x} x^{b-c} = \Gamma(c)/\Gamma(b)$$

and since $2\nu \neq 1$ the two terms inside the bracket behave differently for large a. Hence we can select a such that the last limit is $\neq 0$. Then for a suitable constant $C_0 \neq 0$ we have

$$(94) \qquad |\Phi_{\lambda,m}(\tanh t_m| \geq C_0 |m|^{\frac{1}{2}(\mathrm{Re}(i\lambda)-1)}$$

On the other hand, putting $z = \tanh t_m \, e^{i\theta}$ in (91) we obtain

$$(95) \qquad \sum_n |\Phi_{\lambda,n}(\tanh t_m)|^2 |a_n|^2 = \frac{1}{2\pi} \int_0^{2\pi} |f(\tanh t_m \, e^{i\theta})|^2 d\theta,$$

which by the assumption $f \in \mathcal{E}_\lambda^*(D)$ has a bound $C|m|^b$ for some constants C and b. But (94) and (95) then imply

$$(96) \qquad |a_m|^2 \leq (C/C_0) \; |m|^{b + \frac{1}{2}(1-\mathrm{Re}(i\lambda))}$$

so by Lemma 4.28, T is a distribution and the theorem is proved.

Remark. While the case $i\lambda = 1$ was excluded above the theorem is valid for this case. The proof is even easier. In fact, we have here $\Phi_{\lambda,m}(\tanh t) = \tanh^{|m|} t$ so

$$\lim_{|m| \to \infty} \Phi_{\lambda,m}(\tanh t_m) = e^{-\frac{1}{2}a} .$$

Hence (94) is obvious and (96) follows as before. Thus the ordinary harmonic functions

$$u(z) = \int_B \frac{1-|z|^2}{|z-b|^2} \, dT(b)$$

which are Poisson integrals of distributions are characterized by the growth condition

$$|u(z)| \leq C \, e^{ad(o,z)} .$$

It is of interest to compare this with the classical characterization (Herglotz, Evans, Charathéodory, cf. Nevanlinna [1936]) of Poisson integrals of measures as the harmonic functions u satisfying

$$\sup_{r<1} \int_0^{2\pi} |u(re^{i\phi})| \, d\phi < \infty.$$

§5. The Sphere S^2 Reconsidered.

While our definition and results for the Fourier transform on the hyperbolic plane H^2 have strong analogies with the Euclidean case $\mathbb{R}^2$, this is not so for the theory of spherical harmonics sketched in §3. In fact the orthonormal basis S_{km} of the eigenspace E_k could be chosen entirely at random.

We shall now develop a different kind of Fourier Analysis on S^2; the theory which is due to Sherman [1975] is motivated by the non-Euclidean Fourier analysis in §4.

Let $\mathbf{S}^2$ denote the unit sphere in $\mathbb{R}^3$. Fix a point $a \in S^2$ and let $B = \{s \in \mathbf{S}^2 : (a,s) = 0\}$ (a is the North Pole and B the Equator). In analogy with (11)§4 we consider for $n \in \mathbb{Z}^+$, $b \in B$, the function

$$(1) \qquad e_{n,b}(s) = (a+ib,s)^n, \qquad s \in \mathbf{S}^2,$$

which by Theorem 3.1 belongs to the eigenspace E_n of the Laplacian L for the eigenvalue $-n(n+1)$. (Here $(,)$ denotes the inner product on $\mathbb{R}^3$ extended linearly to $\mathbb{C}^3$.) Along with (1) we consider the function

$$(2) \qquad f_{n,b}(s) = \operatorname{sgn}(s,a)(a+ib,s)^{-n-1}, \quad s \in \mathbf{S}^2-B.$$

Apart from the equator B (where the first factor is undefined and where the second factor has a singularity at two points) $f_{n,b}$ is again an eigenfunction of L for eigenvalue $-n(n+1)$. (Compare with the functions $e_{\mu,b}$, $e_{-\mu+2,b}$ in Lemma 4.1.)

As observed during the proof of Theorem 3.1, the eigenspace E_n contains a unique function invariant under the group $K = \mathbb{O}(2)$ of rotations around a and taking there the value 1. If db is the normalized invariant measure on B this function equals

$$\int_B e_{b,n}(s)\,db = P_n((s,a)), \tag{3}$$

where P_n is a polynomial, the Legendre polynomial of degree n. For the same reason we have

$$\int_B f_{b,n}(s)\,db = P_n((s,a)) \tag{4}$$

for $(s,a) > 0$. For $(s,a) < 0$ the relation holds up to a constant factor which however equals 1 since by (3), $P_n(-1) = (-1)^n$. We shall now prove the analog of formula (47) §4 which was behind the Fourier inversion formula for the hyperbolic plane.

THEOREM 5.1. *Let* $n \in \mathbb{Z}^+$, $s_1 \in S^2$, $s \in S^2-B$. *Then*

$$P_n((s_1,s)) = \int_B e_{b,n}(s_1) f_{b,n}(s)\,db. \tag{5}$$

Proof. Let the integral on the right be denoted

by $I(s_1,s)$. Let $U = \mathbb{O}(3)$ and $\mathfrak{u}$ its Lie algebra which can be identified with the Lie algebra of 3×3 real skew-symmetric matrices. Each $T \in \mathfrak{u}$ gives rise to a right-invariant vector field $\overline{T}$ on U defined by

$$(\overline{T}f)(u) = \left\{ \frac{d}{dt} f(\exp(t\,T)\,u) \right\}_{t=0} . \tag{6}$$

We shall apply this to the function

$$j(u) = I(us_1, us), \qquad u \in U(s),$$

where $U(s) = \{u \in U\colon us \notin B\}$. We shall prove j to be constant.

Let (e_1, e_2, e_3) be the canonical basis of $\mathbb{R}^3$ where $a = e_3$. We put

$$b=b_\theta=e_1 \cos\theta + e_2 \sin\theta \qquad c=c_\theta=-e_1 \sin\theta + e_2 \cos\theta \tag{7}$$

so c is a basis of the tangent space B_b. Given $T \in \mathfrak{u}$ (viewed as a linear transformation of $\mathbb{C}^3$) we denote by β_T the complex vector field on B given by

$$\beta_T(b) = (T(a+ib),c)c. \tag{8}$$

For a fixed $s \in \mathbf{S}^2$ let ϕ be the function on $U \times B$ defined by

$$\phi(u,b) = e_{b,1}(us) = (a+ib,us).$$

LEMMA 5.2. *For each* $n \in \mathbb{Z}$ *we have*

$$-i\ \overline{T}_u(\phi(u,b)^n) = n(Ta,b)\phi(u,b)^n + (\beta_T)_b(\phi(u,b)^n)$$

where the subscripts indicate the variables on which the vector fields $\overline{T}$ *and* β_T *act.*

Proof. We have by (6)

$$\overline{T}_u(\phi(u,b) = \{\frac{d}{dt}\ \phi(\exp t\, Tu,b)\}_{t=0}$$

$$= (a+ib,T(us)) = -(T(a+ib),us).$$

By the skew-symmetry of T, $(Tx,x) = 0$ for $x \in \mathbb{R}^3$ so we have

$$Ta = (Ta,b)b+(Ta,c)c$$

$$Tb = (Tb,a)a+(Tb,c)c = -(Ta,b)a+(Tb,c)c.$$

Using these relations the expression for $\overline{T}_u(\phi(u,b))$ becomes

$$i(Ta,b)\phi(u,b) - (T(a+ib),c)(c,us)$$

$$= i(Ta,b)\phi(u,b) + i(\beta_T)_b(\phi(u,b)),$$

proving the case $n = 1$. More generally,

$$-i\overline{T}\,\phi^n = n\,\phi^{n-1}(-i\overline{T}\phi)$$

$$= n\,\phi^{n-1}\left\{(Ta,b)\phi+\beta_T\phi\right\} = n(Ta,b)\phi^n+\beta_T(\phi^n).$$

LEMMA 5.3. Let $f \in C^1(B)$ and $T \in \mathfrak{U}$. Then

$$\int_B [(Ta,b)f(b) - (\beta_T f)(b)]db = 0.$$

Proof. If (T_{ij}) is the matrix expression for T, i.e. $Te_j = \sum_i T_{ij}e_i$ then by (7), (8),

$$(\beta_T f)(b) = (-\sin\theta\, T_{13} + \cos\theta\, T_{23} + i\, T_{21})\frac{\partial f}{\partial\theta}$$

and

$$\langle Ta,b\rangle f = (\cos\theta\, T_{13} + \sin\theta\, T_{23})f$$

so the lemma follows.

LEMMA 5.4. For each $T \in \mathfrak{U}$, $\overline{T}j = 0$ on $U(s)$.

Proof. For $u \in U(s)$ let

$$\psi(u,b) = e_{b,n}(us_1)f_{b,n}(us).$$

Then by Lemma 5.2,

$$-i\,\overline{T}_u\,(\psi(u,b)) = n(Ta,b)\psi(u,b) + (\beta_T)_b(e_{b,n}(us_1))f_{b,n}(us)$$

$$+\ (-n-1)(Ta,b)\psi(u,b) + e_{b,n}(us_1)(\beta_T)_b(f_{b,n}(us)).$$

$$= -\ (Ta,b)\psi(u,b) + (\beta_T)_b(\psi(u,b)).$$

Using Lemma 5.3 we deduce

$$(\overline{T}j)(u) = \overline{T}_u\left(\int_B \psi(u,b)\,db\right) = \int_B \overline{T}_u(\psi(u,b))\,db = 0.$$

This lemma shows that j is locally constant on $U(s)$. If $(s,a) > 0$ we can find a one parameter group u_t of rotations such that $u_t \in U(s)$ $(0 \leq t \leq 1)$, $u_0 s = s, u_1 s = a$. Then $j(u_0) = j(u_1)$ so

$$\int_B e_{b,n}(s_1)f_{b,n}(s)\,db = \int_B e_{b,n}(u_1 s_1)\,db$$

$$= P_n((u_1 s_1,a)) = P_n((s_1,s)).$$

If $(s,a) < 0$ we use the same argument on $-s$. This proves Theorem 5.1.

We shall now indicate how Fourier analysis on S^2 can be based on Theorem 5.1 in a manner resembling the way the Fourier inversion formula on H^2 was derived from (47) §4. Let $\varphi_n(s) = P_n((s,a))$ $(n \in Z^+)$. Up to a constant factor this is the unique K-invariant function in the eigenspace E_n.

This implies that in analogy with (43) §4 it satisfies the functional equation

$$\int_K \phi_n(uk\cdot s)dk = \phi_n(u\cdot a)\phi_n(s). \tag{9}$$

Let $d_n = (\int |\phi_n(s)|^2 ds)^{-1}$ where ds is the Euclidean surface element on S^2 and let the Haar measure du on U be normalized by

$$\int_U f(u\cdot a)du = \int_{S^2} f(s)ds.$$

If $F \in C^\infty(S^2)$ is K-invariant we have by Theorem 3.1 (iii)

$$F = \sum_0^\infty d_n\, a_n\, \phi_n, \qquad a_n = \langle F, \phi_n \rangle, \tag{10}$$

the series converging in the L^2-norm. Considering the expansion

$$(-L)^p F = \sum_0^\infty d_n (n(n+1))^p a_n\, \phi_n$$

one can prove that (10) converges absolutely and uniformly.

If $f \in C^\infty(S^2)$ we use (10) on the function

$$F(s) = \int_K f(uk\cdot s)dk$$

$u \in U$ being any fixed element and dk being the normalized Haar measure on K. Since ϕ_n is real,

$$\langle F,\phi_n\rangle = \int_{S^2} F(s)\phi_n(s)ds = \int_{S^2}\left(\int_K f(uk\cdot s)dk\right)\phi_n(s)ds$$

$$= \int_U f(uv\cdot a)\phi_n(v\cdot a)dv = \int_U f(uv\cdot a)\phi_n(v^{-1}\cdot a)dv$$

$$= \int_U f(g\cdot a)\phi_n(g^{-1}u\cdot a)dg,$$

which we write as $(f*\phi_n)(u\cdot a)$. Thus we obtain (for $s = a$) the expansion

$$(11) \qquad f = \sum_0^\infty d_n f * \phi_n \;, \qquad f \in C^\infty(S^2).$$

In analogy with (14) §4 we define the <u>Fourier transform</u> of f by

$$(12) \qquad \tilde{f}(b,n) = \int_{S^2} f(s) f_{b,n}(s)ds, \qquad b \in B,\ n \in \mathbb{Z}^+,$$

ignoring the convergence difficulties stemming from the singularities of $f_{b,n}$. Calculating formally, we have

$$\int_B e_{b,n}(s_1)\tilde{f}(b,n)db = \int_{S^2} f(s)\left[\int_B e_{b,n}(s_1)f_{b,n}(s)db\right]ds$$

which by Theorem 5.1 reduces to

$$\int_{S^2} f(s)P_n((s_1,s))ds = \int_{S^2} f(u\cdot a)P_n((s_1,ua))du.$$

Thus we have the compact analog of Lemma 4.12

(13) $$(f*\phi_n)(s) = \int_B e_{b,n}(s)\tilde{f}(b,n)\,db.$$

Combining this with (11) we are led to the following result.

THEOREM 5.5. *For* $f \in C^\infty(S^2)$ *define the Fourier transform by* (12). *Then*

(14) $$f(s) = \int_{B\times\mathbb{Z}^+} e_{b,n}(s)\tilde{f}(b,n)\,d(b,n),$$

where $d(b,n)$ is the measure $d_n db$ on $B \times \mathbb{Z}^+$.

For a complete proof of this result, including the precise definition of (12) see Sherman [1975]. There is also an alternative version to (12) and (14) with $e_{b,n}$ and $f_{b,n}$ interchanged.

BIBLIOGRAPHICAL NOTES

§2. An integral formula for the eigenfunctions of $L_{\mathbb{R}^n}$ was given by Hashizume et al. [1972]. The more specific version in Theorem 2.1 was proved by the author [1974] and was extended to higher dimensions by Morimoto [1]. Different integral representations are given by Ehrenpreis [1969], Ch. VII and Palamodov [1970], Ch. VI. Lemma 2.2 is an immediate generalization of a classical result on entire functions of exponential type, (Boas [1954]). Propositions 2.5 and 4.26 are special cases of a mean value theorem for homogeneous spaces (cf. Helgason [1962], Ch. X, Cor. 7.4). The eigenspace irreducibility (Theorem 2.6 and Prop. 2.9) is from the authors' papers [1974], [1977]. The higher dimensional versions require some modification. The extension of Theorem 2.6 to symmetric spaces was given by the author [1980]. The Paley-Wiener type theorem (Theorem 2.10) is from Helgason [1976], §11. (The theorem is misstated in this paper: "polynomial" in Theorem 11.1 should read "harmonic polynomial").

§3. Theorem 3.1 is classical. See Cartan [1929] for the extension of Part (ii) to symmetric spaces. For generalizations of (4) and (5) see Maass [1959], Helgason [1963], and Kostant-Rallis [1971].

§4. No.1. Theorem 4.2 is a special case of results from the author's papers [1965], [1970], Ch. III, §5 and [1973] §8 for general symmetric spaces. Another proof of Part (ii) was given by Torasso [1977]. Theorem 4.3 was proved by the author ([1970], Ch. IV, §1 and [1974]). It was extended to Riemannian symmetric spaces by Kashiwara et al. [1978] and to some non-Riemannian symmetric spaces by Oshima-Sekiguchi [1980]. Related results are given in Korányi-Malliavin [1975], Johnson-Korányi [1980] and Berline-Vergne [1980].

The irreducibility criterion for the eigenspace representation (Theorem 4.4) was proved by the author [1970], Ch. IV and extended in [1970], [1976] to symmetric spaces.

§4, No.2. Spherical functions on symmetric spaces go back to Cartan [1929], Gelfand-Naimark [1950] and Gelfand [1950]. For general symmetric spaces the integral formula (19) and the expansion in Theorem 4.6 were given by Harish-Chandra [1954], [1958]. Theorem 4.7 (of Paley-Wiener type) was proved in a different way by Ehrenpreis-Mautner [1955]. The proof in the text is from the author's paper [1966] where the result is extended to symmetric spaces except for the term-by-term integration $\int \Sigma_n = \Sigma_n \int$ unjustified. Gangolli's paper [1971] gave the required justification which was not a routine matter. (See Helgason [1970], p.38-39 for a simplified justification). For the rank-one case other proofs (with generalizations) are given by Flensted-Jensen [1972], Koornwinder [1975] and Chèbli [1974].

In the text we have given two proofs of Theorem 4.8 (the Plancherel theorem). The former proof is the one given by Rosenberg [1977] and along with the generalization of Theorem 4.7 mentioned above it extends to symmetric spaces. The latter proof of Theorem 4.8 is from Godement [1957] where the inversion formula is related to that of Fock-Mehler transform. An interesting interpretation and generalization of the formula for $\int (\sin \lambda t) \operatorname{sh}(t)^{-1} dt$ was discovered by Flensted-Jensen [1978]; it leads to significant simplifications when G is a normal form. The more general Plancherel theorem for the group $G = SL(2,\mathbb{R})$ itself was already given by Harish-Chandra [1952], whereas the basic representation theory was developed by Bargmann [1947].

A measure-theoretic Plancherel formula for the spherical transform on a symmetric space was given by Mautner [1951]. A more precise topological version was given by Godement [1951], [1957]. In contrast to these, Harish-Chandra gives the formula ([1958 I], [1958 II], [1966]) with an explicit description of the dual measure in terms of the c-function. An important contribution to the proof was given by Gindikin-Karpelevič [1962].

<u>§4, No.3.</u> A generalization of the integral formula in Theorem 4.16 to symmetric spaces was proved by the author [1976], Cor. 7.4; this paper also gives a simplified proof of Harish-Chandra's expansion for $\Phi_{\lambda,m}$ (Theorem 4.18).

The Schwarz theorem was extended to symmetric spaces by Karpelevič [1965]; see also Lowdenslager [1958] and Moore [1964] for the case of bounded symmetric domains.

In the Mathematical Reviews article on Furstenberg [1963] there was posed the problem of generalizing Fatou's theorem (Theorem 4.21) to symmetric spaces X of the noncompact type. The theorem sought was the following:

<u>Let</u> u <u>be a bounded solution of Laplace's equation</u> $Lu = 0$ <u>on</u> X <u>and let</u> $p \in X$ <u>be any point</u>. <u>Then for</u>

almost all geodesics γ *starting at* p

$$\lim_{t \to \infty} u(\gamma(t)) \quad \textit{exists.}$$

This "*Fatou theorem for symmetric spaces*" was proved by Helgason-Korányi [1968]. This result has had many generalizations and analogs proved by Knapp, Williamson, Korányi, Lindahl, Stein, N. Weiss, Michelson, Urakawa, Linden, Putz and Taylor; see Korányi [1970], [1979] for a detailed survey.

For symmetric spaces the limit theorem (Theorem 4.27) was proved by Harish-Chandra [1958 I] p.291 for the case when the hyperfunction T is constant and by the author [1970] p.129 in general. The proof in the text is due to Lewis [1970].

Theorem 4.29 is due to Lewis [1978], as well as its extension to symmetric spaces of rank one. Lewis proved also for symmetric spaces of any rank that the Poisson transform $P_\lambda: T \longrightarrow f$ satisfies $P_\lambda(\mathcal{D}'(B)) \subset \mathcal{E}^*_\lambda(D)$ for each λ. The converse inclusion was proved by Oshima-Sekiguchi [1980] (for λ outside some hyperplanes).

§5. The results here are due to Sherman [1975]. The extension to compact symmetric spaces of rank one and in part to such spaces of any rank is outlined in Sherman [1977].

REFERENCES

Bargmann, V. [1947], Irreducible unitary representations of the Lorentz group, Ann. of Math. 48(1947), 568-640.

Berline, N. and M. Vergne, [1980] Équations de Hua et intégrales de Poisson, C.R. Acad. Sci. Paris 290 (1980) A, 123-125.

Boas, R.P. Jr. [1954], Entire Functions. Academic Press, New York, 1954.

Bochner, S. [1955], Harmonic Analysis and the Theory of Probability. Univ. of Calif. Press, Berkely, 1955.

Cartan, É. [1929], Sur la détermination d'un système orthogonal complet dans un espace de Riemann symétrique clos. Rend. Cir. Mat. Palermo 53 (1929), 217-252.

Chèbli, H. [1974], Sur un théorème de Paley-Wiener associe à la décomposition spectrale d'un opérateur de Sturm-Liouville sur (o,∞). J. Funct. Anal. 17(1974), 447-461.

Dym H. and H.P. McKean, [1972], Fourier Series and Integrals, Academic Press, 1972.

Ehrenpreis, L. [1969], Fourier Analysis in Several Complex Variables, Wiley, New York, 1969.

Ehrenpreis, L. and F.I. Mautner, [1955], Some properties of the Fourier transform on semisimple Lie groups. I. Ann. of Math. 61(1955), 406-439.

Erdélyi, A. et al., [1953] Higher Trancendental Functions (Bateman Manuscript Project) Vol. I. McGraw-Hill Book Co. New York, 1953.

Eymard, P. [1977], Le noyau de Poisson et la théorie des groupes. Symposia Mathematica 1977.

Flensted-Jensen, M. [1972], Paley-Wiener type theorems for a differential operator connected with symmetric spaces. Ark. för Mat. 10(1972), 143-162.

Flensted-Jensen, M. and T. Koornwinder [1973], The convolution structure for Jacobi function expansions. Ark. för Mat. 11(1973), 245-262.

Flensted-Jensen, M. [1978], Spherical functions on a real semisimple Lie group. A method of reduction to the complex case. J. Funct. Anal. 30(1978), 106-146.

Furstenberg, H. [1963], A Poisson formula for semisimple Lie groups, Ann. of Math. 77(1963), 335-386.

Gangolli, R. [1971], On the Plancherel formula and the Paley-Wiener theorem for spherical functions on semisimple Lie groups. Ann. of Math. 93(1971), 150-165.

Gelfand, I.M. [1950], Spherical functions on symmetric spaces. Dokl. Akad. Nauk.SSSR 70(1950), 5-8.

Gelfand, I.M. and M.A. Naimark [1950], Unitary Representations of the Classical Groups. 1950, German transl. Akademie-Verlag, Berlin 1957.

Gelfand, I.M. and M.I. Graev and N.Ya. Vilenkin [1966], Generalized Functions Vol. 5. Academic Press, New York 1966.

Gindikin, S.G. and F.I. Karpelevič [1962], Plancherel measure of Riemannian symmetric spaces of non-positive curvature. Sov. Math. 3(1962), 962-965.

Godement, R. [1951], Sur la théorie des représentations unitaries. Ann. of Math. 53(1951), 68-124.
[1957], Introduction aux travaux de A. Selberg. Séminaire Bourbaki 1957.

Harish-Chandra, [1952], Plancherel formula for the 2 × 2 real unimodular group. Proc. Nat. Acad. Sci. USA 38(1952), 337-342.
[1954], Representation of semisimple Lie groups II. Trans. Amer. Math. Soc. 76(1954), 26-65.
[1958 I], Spherical functions on a semisimple Lie group I. Amer. J. Math. 80(1958), 241-310.
[1958 II], Spherical functions on a semisimple Lie group II. Amer. J. Math. 80(1958), 553-613.
[1966], Discrete series for semisimple Lie groups II. Acta. Math. 116(1966), 1-111.

Hashizume, M.A. Kowata, K. Minemura, and K. Okamoto [1972], An integral representation of an eigenfunction of the Laplacian on the Euclidean space. Hiroshima Math. J. 2(1972), 535-545.

Helgason, S. [1962], Differential Geometry and Symmetric Spaces. Academic Press, New York, 1962.
[1963], Invariants and fundamental functions. Acta Math. 109(1963), 241-258.
[1965], Radon-Fourier transforms on symmetric spaces and related group representations. Bull. Amer. Math. Soc. 71(1965), 757-763.
[1966], An analog of the Paley-Wiener theorem for the Fourier transform on certain symmetric spaces. Math. Ann. 165(1966), 297-308.
[1970], A duality for symmetric spaces with applications to group representations. Advan. math. 5(1970), 1-154.

[1973], The surjectivity of invariant differential operators on symmetric spaces I. Ann. of Math. 98(1973), 451-479.

[1974], Eigenspaces of the Laplacian; integral representations and irreducibility. J. Functional Analysis 17(1974), 328-353.

[1976], A duality for symmetric spaces with applications to group represention*s*. II. Differential equations and eigenspace representations. Advan. Math. 22(1976), 187-219.

[1977], Some results on eigenfunctions on symmetric spaces and eigenspace representations. Math. Scand. 41(1977), 79-89.

[1979], Invariant differential operators and eigenspace representations pp.236-286 in Atiyah et. al. Representation theory of Lie groups. London Math. Soc. Lecture Notes 34, Cambr. Univ. Press. 1979.

[1980], A duality for symmetric spaces with applications to group representations. III. Tangent space analysis. Advan. math. 36(1980), 297-323.

Helgason S. and A. Korányi [1968], a Fatou-type theorem for harmonic functions on symmetric spaces. Bull. Amer. Math. Soc. 74(1968), 258-263.

Hörmander, L. [1963], Linear Partial Differential Operators. Springer-Verlag 1963.

Johnson, K. and A. Korányi [1980], The Hua operators on bounded symmetric domains of tube type. Ann. of Math. 111(1980), 589-608.

Karpelevič, F.I. [1965], The geometry of geodesics and the eigenfunctions of the Beltrami-Laplace operator on symmetric spaces. Trans. Moscow Math. Soc. 14(1965), 51-199.

Kashiwara, M., A. Kowata, K. Minemura, K. Okamoto, T. Oshima and M. Tanaka, [1978], Eigenfunctions of invariant differential operators on a symmetric space. Ann. of Math. 107(1978), 1-39.

Koornwinder, T.H. [1975], A new proof of a Paley-Wiener theorem for the Jacobi transform. Ark. för. Mat. 13(1975), 145-159.

Korányi A. [1970], Generalizations of Fatou's theorem to symmetric spaces. Rice Univ. Studies 56(1970), 127-136.

[1979], A survey of harmonic functions on symmetric spaces. Proc. Symp. Pure Math. Vol. XXXV Part 1, Harmonic Analysis in Euclidean Spaces, Amer. Math. Soc. Providence 1979.

Korányi, A. and P. Malliavin, [1975], Poisson formula and compound diffusion associated to an overdetermined elliptic system on the Siegel halfplane of rank two. Acta. Math. 134(1975), 185-209.

Kostant, B. and S. Rallis, [1971], Orbits and Lie group representations associated to symmetric spaces. Amer. J. Math. 93(1971), 753-809.

Lang. S. [1975], $SL_2(\mathbb{R})$. Addison-Wesley, Reading, Mass. 1975.

Lewis, J.B. [1970], Eisenstein series on the boundary of the disk. Thesis, MIT, 1970.

[1978], Eigenfunctions on symmetric spaces with distribution-valued boundary forms. J. Funct. Analysis, 29(1978), 287-307.

Lowdenslager, D.B. [1958], Potential theory in bounded symmetric homogeneous complex domains. Ann. of

Math. 67(1958), 467-484.

Maass, H. [1959], Zur Theorie der harmonischen Formen Math. Ann. 137(1959), 142-149.

Magnus W., R. Oberhettinger, [1948], Formeln und Sätze für die speciellen Funktionen der mathematischen Physik. Springer, Berlin-Göttingen-Heidelberg, 1948.

Mautner, F.I. [1951], Fourier analysis and symmetric spaces. Proc. Nat. Acad. Sci. USA 37(1951), 529-532.

Moore, C.C. [1964], Compactifications of symmetric spaces II; The Cartan domains. Amer. J. Math. 86(1964), 358-378.

Morimoto, M. [1], Analytic functionals on the sphere and their Fourier-Borel transformations. Banach Center Publications. (to appear).

Nevanlinna, R. [1936], Eindeutige analytische Funktionen. Springer Verlag, Berlin, 1936.

Oshima, T. and J. Sekiguchi, [1980], Eigenspaces of invariant differential operators on an affine symmetric space. Invent. Math. 57(1980), 1-81.

Palamodov, V.P. [1970], Linear Differential Operators with Constant Coefficients. Springer, New York 1970.

Rosenberg, J. [1977], A quick proof of Harish-Chandra's Plancherel theorem for spherical functions on a semisimple Lie group. Proc. Amer. Math. Soc. 63 (1977), 143-149.

Schempp W. and B. Dreseler, [1980], Einführung in die Harmonischc Analyse. B.G. Teubner, Stuttgart, 1980.

Schwartz, L. [1966], Théorie des Distributions. Hermann and Co., Paris, 1966.

Sherman, T.O. [1975], Fourier analysis on the sphere. Trans. Amer. Math. Soc. 209(1975), 1-31.
[1977], Fourier analysis on compact symmetric space. Bull. Amer. math. Soc. 83(1977), 378-380.

Sugiura, M. [1975], Unitary Representations and Harmonic Analysis. Halsted Press. New York 1975.

Takahashi, R. [1963], Sur les représentations unitaires des groupes de Lorentz généralisees. Bull. Soc. Math. France 91(1963), 289-433.

Torasso, P. [1977], Le théorème de Paley-Wiener ..., J. Funct. Anal. 26(1977), 201-213.

Trombi, P. and V. Varadarajan [1971], Spherical transform on semisimple Lie groups. Ann. of Math. 94(1971), 246-303.

Varadarajan, V.S. [1977], Harmonic Analysis on Real Reductive Groups, Lecture Notes in Mathematics Springer, New York 1977.

Vilenkin, N. Va. [1968], Special Functions and the Theory of Group Representations. Transl. of Math. Monogr. Amer. Math. Soc. 22(1968).

Wallach, N.R. [1973], Harmonic Analysis on Homogeneous Spaces. M. Dekker, New York, 1973.

Warner, G. [1972], Harmonic Analysis on Semisimple Lie Groups I, II. Springer, Berlin-Heidelberg-New York, 1972.

Weil, A. [1940], L'Intégration dans les Groupes Topologiques et les Applications. Hermann, Paris, 1940.

Zariski, O, and P. Samuel, [1960], Commutative Algebra, Vol. II. Van Nostrand, Princeton, N.J., 1960.

SUBJECT INDEX

Date Due
